A Scientific Approach to Scientific Writing

John Blackwell · Jan Martin

A Scientific Approach to Scientific Writing

Springer

John Blackwell
Sees-editing Ltd
Weston-super-mare, BS23 2JU, UK
jrb@sees-editing.co.uk

Jan Martin
Sayer-Martin Ltd
Aberaeron, SA46 0HS, UK
drjan.martin@virgin.net

ISBN 978-1-4419-9787-6 e-ISBN 978-1-4419-9788-3
DOI 10.1007/978-1-4419-9788-3
Springer New York Dordrecht Heidelberg London

Library of Congress Control Number: 2011926358

Printed on acid-free paper

Springer is part of Springer Science+Business Media (www.springer.com)

Acknowledgements

This book would probably not have been written without suggestions and encouragement by Jianquan Liu (Professor of Molecular Ecology, Lanzhou University, China) and Xuxia Zhao (International Coordinator, Northwest Institute of Plateau Biology, Chinese Academy of Sciences). In addition, valuable comments have been made at various points in its writing by Dr. David Blackwell (Organic Chemistry, Cambridge), Elena Conti (Professor of Systematic Botany, University of Zürich, Switzerland), Prof. Anna Linusson (Department of Chemistry, Umeå University, Sweden), Prof. Thomas Moritz (Head of Department of Forest Genetics and Plant Physiology, SLU, Sweden), Prof. Michal Otyepka (Head of Department of Physical Chemistry, Palacky University, Czech Republic), Prof. Göran Sandberg (Executive Director, Knut and Alice Wallenberg Foundation, Sweden), Dr. Johan Sandberg (Department of Informatics, Umeå University, Sweden), Prof. Ian Scott (IBERS, Aberystwyth University, UK), Prof. Bo Strandberg (Sahlgrenska Academy at University of Gothenburg), Dr. Jan-Olov Weslien (Skogforsk, Sweden) and Prof. Mike Young (IBERS, Aberystwyth University, UK).

Thanks are also due to Joëlle Hoggan, Jon Sayer and Eveline Forrest (Forrest Text) for their help at various stages of writing this book.

Contents

Chapter 1
Maximizing Chances of Publication

In the modern world, every scientist who wants to publish findings in an international, peer-reviewed journal must write in English. This can be very challenging for people who are not native speakers of English. Indeed, it can be challenging for people who *are* native speakers. However, whether you are writing papers in your first or any other language, the process can be greatly facilitated by approaching it in a logical, systematic manner.

The authors of this guide have written, re-written or edited more than 4,000 texts of diverse kinds and (together with numerous friends, colleagues and clients who have commented on various drafts) have substantial experience of both writing papers and the problems encountered by people who are learning to write them, in either their first or other languages. This brief guide is intended to help people tackle these problems, using four increasingly complex hypothetical case studies. The book focuses on writing research papers because they are key outputs for scientists seeking a high-profile career and the main elements of the primary literature. However, as discussed in Chapter 7, the outlined systematic approach can be applied when writing other kinds of presentations (e.g., reports, reviews and oral presentations).

It is impossible to describe how to write anything without considering some linguistic aspects. Here, we address aspects we believe to be critical for structuring a paper, but it is assumed that readers will have had many years of English education; hence attempting to add significantly to your knowledge of the language would be pointless. Furthermore, there are many good books on English grammar, for people of all ages. We are not intending to add to their number. Instead, the main aim is *to show you ways to arrange, compose and present a study*, starting from brief, simple statements, in order to *maximize the chances of publication*.

There are several ways to do this. The most effective way of all is to buy the company that owns the target journal. Then you can publish all of your papers, and those of your friends and colleagues, in it. The problems with this strategy are that it is very expensive and there is a risk that the quality, and hence the impact rating, of the journal will decline. The second strategy is to discover an embarrassing secret about the journal's editor and apply blackmail. The problems with this approach are that it is difficult and immoral. The third option is to describe the problems/phenomena

J. Blackwell, J. Martin, *A Scientific Approach to Scientific Writing*,
DOI 10.1007/978-1-4419-9788-3_1, © Springer Science+Business Media, LLC 2011

you have addressed, your results and their implications clearly and concisely. This is the approach that can be most easily adopted by the majority of readers and the one that will be considered here.

As discussed in Chapter 6, however brilliant or innovative your study may be, there is no guarantee that it will be accepted by the target journal. However, presenting your study clearly and coherently greatly improves the chances of acceptance. Furthermore, *writing papers is simple*. Essentially, all that you need to do is this: Describe what you have done and why you have done it, outline the results and limitations (or *focus*) of the study, discuss the implications of the findings and highlight their importance.

The problem is that this has to be done:

- clearly and concisely
- while covering all the key points and showing that your study is very important
- and in a foreign language, if you are writing in English and you are not a native English speaker.

As noted before, currently, if one wants to publish papers in international, peer-reviewed journals, they must be written in English. *This is clearly unfair* and may change in the future. Indeed, as China's economic power grows, it is quite possible that in a few decades everyone will have to write in Chinese. However, at the moment all scientists who want a successful career have to write in English. Thus, the objective of this book is to present a coherent, systematic strategy that can help both native and non-native English speakers to construct, write and publish papers, and other kinds of scientific communications, in English.

Chapter 2
Essential Steps Before Writing a Paper

Having completed a study and acquired all the data required to present it, you are ready to begin preparing a paper. However, before beginning to write, you have to take several critical preliminary steps: Your notes must be gathered, a suitable place for writing must be found, a selective literature review may be helpful, a target journal must be identified, linguistic limitations must be recognized, the study must be defined and delimited, and the information must be arranged. Failure to take these steps will make writing more difficult and seriously compromise the chances of publication. Therefore, this chapter outlines what needs to be done in each of these steps.

2.1 Gather Your Notes

Writing a paper can be greatly facilitated by keeping good notes during the study phase. A comprehensive laboratory or field notebook (either electronic or handwritten) is a valuable resource when you finally sit down to write your paper. It is easy, during data collection, to assume that you will be able to remember why or how you did things. Unfortunately, when you begin to write up your work, several months or even years later, you are likely to find it difficult to recall details. Keeping thorough notes as you proceed, supported by other sources of information such as sketches or photographs, can alleviate a great deal of stress in the latter stages of your study. Of course, it is not only essential to keep good notes, it is also essential to use them. Thus, before sitting down to write, all the relevant notes must be gathered.

2.2 Find Somewhere Quiet

Having gathered your notes, it is essential to find somewhere peaceful to write, or at least somewhere where disturbances are minimal, because writing well requires intense concentration. Ideally, one should find a quiet room and pin a 'Do not disturb' note on the door while arranging the material and writing. Supervisors can help by refraining from demanding to see how the work is progressing every few

J. Blackwell, J. Martin, *A Scientific Approach to Scientific Writing*,
DOI 10.1007/978-1-4419-9788-3_2, © Springer Science+Business Media, LLC 2011

minutes. Supervisors have many ways to make students' and post-doctoral workers' lives miserable if they wish, but constantly disturbing them when they are writing is particularly unhelpful; it reduces the quality of the output, thus impairing the chances of publication. On the other hand, an inexperienced writer will need help. Hence, setting a good balance between providing helpful advice and interfering too much is an important 'soft skill'.

2.3 *Selectively* Review the Literature

It is vital to review the relevant literature, to ensure that no important observations that either support your findings or contradict them have been missed. Failure to mention such references will create a poor impression and may seriously impair the chances of publication (especially if you miss publications by one or more of the referees). Thus, it is essential to search all of the relevant databases, such as CAB Abstracts (http://cababstracts.edina.ac.uk) and ISI Web of Knowledge^{SM} (http://apps.isiknowledge.com), using all of the potentially relevant keywords. The websites of these databases themselves, and a number of books and other web sources (e.g., Harvard College Library's site; http://hcl.harvard.edu/research/guides/citationindex/ accessed September 5, 2010), describe ways to search these databases.

However, you can only tell what fields of literature you need to scan when you know precisely what each section of a planned paper is going to cover, and thus the *kinds* of references you need to cite. Otherwise much time could be wasted reading material that is only tangentially related to your study. For instance, if you tried to read every report that has ever been written on over-fertilization of every type of soil, supporting every crop, and all the associated problems, you would never stop reading.

Furthermore, a detailed review of the literature before drafting a paper can be counter-productive, since nearly all of the papers that *could* be read will have some sections that are relevant to your study, but also many sections that have little relevance. Reading such papers can be seriously distracting because they will present many ideas that you may start to think should perhaps be mentioned, complicating rather than helping attempts to draft a clear, coherent framework. In addition, you are likely to be an expert in the subject (otherwise you would not have been able to plan and execute the work) and you should already have good knowledge of the pertinent literature. Therefore, it is generally better to draft your paper first, and *then* identify the aspects of the literature that you can *selectively* focus on. Hence, we recommend restricting any literature survey, at this stage, to at most a couple of recent reviews to refresh your memory about key aspects of the subject that may need to be covered, jotting down brief details of references that *could* be cited. However, even this is optional until the framework of the paper has been drafted.

Ways of identifying references that *need* to be cited while drafting the framework are described in detail later, but here we will mention that key steps in writing several sections of a paper (especially the *Introduction* and *Discussion*) are identifying appropriate references and deciding where they should be placed. There are two classes of references: essential and illustrative. Essential references are those that have to be cited because they are critical for justifying your study, those for instance showing that a model you used provides robust predictions for analyses such as yours. Illustrative references are those that have been selected from a large number that could have been cited, showing (for instance) that over-fertilization of soil can cause poor root development. A systematic way in which both classes of references can be identified and allocated suitable places for citation is shown in the descriptions of procedures applied to construct sections of papers describing the case studies. However, it should be noted here that, if possible, for illustrative references it is sensible to choose those published by likely referees.

2.4 Identify a Target Journal

It is also essential to identify an appropriate target journal. There are several factors to consider here, including the significance of the study, the subject matter and the impact ratings of candidate journals. Assessing the significance of the study is the most difficult, since it is highly subjective. Clearly, all studies are important to the researchers involved, and they are often surprised when friends, relatives and referees fail to see their importance. However, their significance can be roughly assessed by considering the applicability, novelty and generality of the results.

If the results of a study can be applied in multiple disciplines, or major industrial processes, the interest in them will be very high. Similarly, if they include highly unexpected or novel results that are likely to create a major shift in theoretical understanding, the interest will be very high and wide. In addition, if a study has been very extensive, covering large numbers of factors, there is likely to be much greater interest than if the study has been very restricted. In such cases submission to a very highly ranked general science journal, such as *Nature* or *Science*, can be considered. In other cases submission to a journal covering your field of interest is more likely to be successful. It is easy to list possible options in this respect, and both their impact factors and the specific areas that they tend to focus on.

The next step is to identify the journal, amongst likely candidates, with the highest impact factor that routinely accepts papers of similar significance to your study. It is then *essential* to read the journal's instructions for authors *thoroughly*. Astonishingly large proportions of authors either fail to do this, or read them but fail to follow them. Editors of journals find this extremely irritating, since it means that if they accept papers by these authors, a lot of time will have to be wasted telling the authors to amend their papers in accordance with the guidelines. In practice, of course, the editor may simply decide to reject the papers and publish papers by authors who have followed the guidelines instead. Hence, failure to follow the guidelines can seriously compromise the chances of publication (and at best create unnecessary delays).

The journal's guidelines are usually available on-line and should be consulted prior to writing. Perhaps the most obvious restriction is the number of words – it is remarkable how many papers are rejected or require revision because they are too long. Usually the maximum numbers of words allowed for both the *Abstract* and the paper as a whole will be specified. It is wise to know these limits before beginning to write. Indeed, they provide a useful guide. If the maximum length of a paper's main text is 5,000 words, the journal's editor will probably expect most papers to have ca. 4,000–5,000 words. Thus, if your study can be fully covered in less than 2,000 words, there is a substantial chance that the editor will regard it as too slight, that is, as containing too little information (unless the findings are unusually important or unexpected). In such cases, submission to other journals may be more fruitful – or submission to the same journal as a *Short communication*, if it has a section for such contributions.

When you have completed, or nearly completed, a draft of a paper you may decide that the first identified journal is not the best choice, perhaps because the paper is too short, too slight or even possibly that a higher-impact journal might accept it. In such cases another target journal should be identified, and the paper should be adjusted in accordance with that journal's instructions. This is tedious, but it is far better than either sending it to a journal that is likely to reject it or sending it in an inappropriate format.

In addition, there may be limitations on the number of tables or figures and the nature of figures. For example, some publishers ask authors to cover the cost of reproducing figures in color; such expense can be avoided by ensuring that figures are clear in black and white and perhaps providing a link to a website where more detailed versions can be viewed in full color. Furthermore, journals often require either British or US English spellings and grammar to be used. Thus, it is important to use the appropriate language setting and apply your word processor's spell-checking function before submitting a paper. They may also have certain other linguistic requirements, some of which are discussed at various points in the text later and should be followed (if possible).

2.5 Awareness of Linguistic Limitations

Having compiled your data, found a quiet place to work, delimited your study, identified a suitable target journal and thoroughly read its *Instructions for authors*, you are ready to begin writing. However, while switching on your word processor, and throughout the rest of the process, it is important to note that if you are writing in English and you are not a native speaker or highly skilled non-native speaker, your writing style has to be adjusted accordingly. Notably:

- It is essential to write more simply than in your first language.
- Much of the advice in standard textbooks about writing papers in English is not helpful, because it tells you *what* to do, but not *how* to do it, thus it is like a

sculptor saying that to create a model of Napoleon you should form a mental image of him and then mold your material into a likeness of it. This is true, but most of us need a little more guidance.

- It may be impossible for a non-native speaker to cover all the key points *and* be clear *and* concise *and* highlight the importance of the study.

If your language skills are not sufficient to address all of the major points clearly and concisely, it is essential to prioritize *clarity*. Then an English-speaking editor, friend or colleague can understand the paper and if necessary improve the English. To illustrate this point, both authors of this guide can write simple sentences in French, but if we try to write complex sentences they become incomprehensible. Editors and referees have similar problems with many papers written by non-native speakers.

For example, a paper we edited recently was full of sentences like:

> Physical obstacles have been created purely historically reasons and therefore they would be disappeared, especially at nuclear loci quite quickly, but environmental obstacles will probably be persistent for much longer times due to ecological reasons.

Many referees or editors of journals may decide that such papers have to be re-written, because they are too difficult to understand, or merely reject them, because they think that making them comprehensible will require too much effort. Therefore, if you cannot write fluently in English (or any other stipulated language), it is essential to write simply, preferably in short sentences that are easy to understand. An experienced editor can then make the phrasing more elegant. For example, another paper we handled recently had many passages such as:

> A linear correlation between nickel uptake and nitrate uptake was found; the nickel uptake increased with nitrate uptake. Another correlation was found with phosphate uptake; when nickel uptake decreased phosphate uptake also decreased.

This can be stated much more elegantly, as follows: *Positive linear correlations were found between nickel uptake and both nitrate and phosphate uptake.* However, although the text was too long and repetitive, it was easy to understand and edit because the sentences were simple. Thus, it was preferable in many ways to papers with confusingly complex sentences.

Linguistic limitations are further factors that may be considered when choosing a target journal since there are substantial variations in the linguistic standards of journals, and these do not always correlate strongly with their impact factors (i.e., some highly ranked journals do not require the language to be as polished as some less highly ranked journals). Therefore, if you are not a skilled writer, it may be worth identifying journals that sometimes publish papers with poor linguistic standards, especially if you need to publish a paper quickly (for instance to support a grant application). A native English-speaking friend working in your field may be able to help identify such journals.

2.6 Defining and Delimiting 'the Study'

The word 'study' can be confusing because it has several meanings, *inter alia* an investigation of certain phenomena, and a written report of such an investigation. However, for simplicity, in this guide 'a study' always refers to an investigation, and a manuscript describing an investigation is referred to as 'a paper'. Clearly, before starting to write a paper describing a study, it is essential to decide what the paper is going to cover, that is, the study must be delimited. Sometimes this is easy. For instance, a study could be summarized as follows:

- *Rationale*: It is generally believed that mature bananas are yellow and bent. However, the Learned Society of Unorthodox Thinkers (a fictitious body) has recently postulated that they are in fact red and straight, and if they aren't they certainly should be.
- *Objectives*: To test the general belief and the Learned Society's conflicting hypothesis.
- *What was done*: Two thousand mature bananas were acquired and examined.
- *Findings*: All of the mature bananas examined were yellow and bent.
- *Implications*: The results indicate that bananas are generally yellow and bent, supporting the traditional belief (although it is possible that some are red and straight, since the survey was far from comprehensive). Whether they *should* be red and straight requires further investigation.

In this case, delimiting the study is very straightforward. It is also often straightforward in other cases where one or two simple hypotheses are postulated and tested. However, it is not generally quite so easy, because most investigations are much more complex. For example, in PhD projects multiple phenomena are often investigated, which could be reported in (say) three long papers, or larger numbers of short papers. Clearly, in such cases it is essential to decide which parts of the project are going to be covered in a particular paper. Fortunately, researchers usually have intimate knowledge of the scope (and linguistic style) of papers published in journals covering their fields of interest, and this can provide a good guide for deciding how much information should be included in each paper and thus dividing the project into a series of studies.

In addition, the elements of a larger project can usually be grouped into a set of reasonably discrete investigations, which greatly facilitates the delimitation of studies. For example, let us consider the following hypothetical project. The small, fictitious country Sucrosia has a near-ideal location and conditions for producing sugar from sugarcane, hence sugarcane is cultivated in large areas of the country, after which it is milled and the resulting sugar is refined for export. Some of the waste biomass ('bagasse') is also used for cogenerating energy. However, the yields and profitability are generally low by international standards. There are grounds for thinking that the poor yields are partly due to over-fertilization. Thus, this possibility clearly needs to be addressed, but many other variables also need to be considered, including the irrigation strategy applied, the cultivars used, harvesting operations

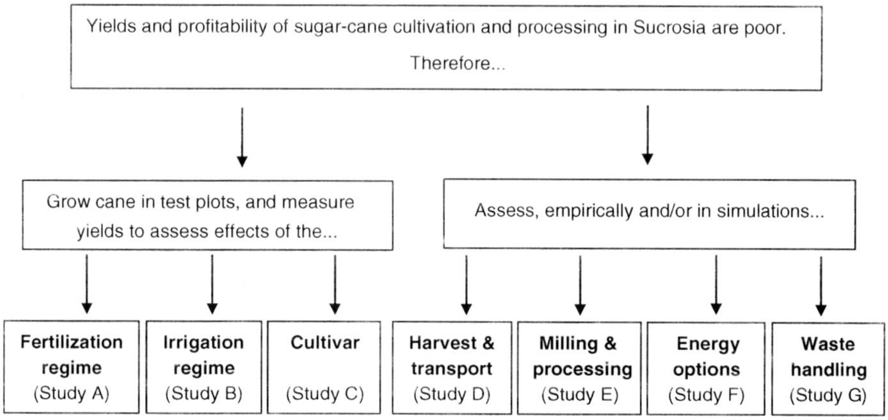

Fig. 2.1 Flow chart delimiting studies within the hypothetical Sucrosia sugarcane project

and scheduling, field-to-mill transport, cut-to-milling delays, the milling and other processing equipment, the use of steam and power, energy cogeneration options and waste treatment. In such cases, as illustrated in Fig. 2.1, flow charts may be very useful for visualizing the work to be done, delimiting studies within the project (and assigning human and other resources to them), tracking progress and (eventually) writing sections of papers and/or reports.

Sub-dividing a project in this manner can also provide a very convenient means for constructing sub-titles of sections of interim and final reports, by simply slightly re-wording the summarizing terms (shown in bold) for each of the delimited studies (e.g., *Optimization of the fertilization regime, Optimization of the irrigation regime* and *Milling and processing strategies*). Of course, each of these studies may be quite complex, so further, more detailed flow charts for each of the studies may be useful, as illustrated for two complex hypothetical case studies in Chapter 4. Such detailed flow charts can provide templates for writing sections of papers (Chapter 4), and can be useful for identifying references that need to be cited in each section, as described later. It should also be noted that there will be substantial overlap between some sections of papers describing these studies, for instance, plots at the same study sites will probably be used for the fertilization, irrigation and cultivar selection studies, so the descriptions of the sites, and the criteria used to select them, will be the same in Studies A, B and C. This is convenient, because these aspects of the studies need to be described in detail only once, and after (say) writing a paper on optimization of the fertilization regime, papers on optimization of the irrigation regime and cultivar selection can refer to information in the first paper.

Having divided a project into discrete studies that can be described in papers of appropriate length, a related problem is deciding where to start from, that is, what aspect of each study to describe first. Some authors recommend starting by describing *what was done*, that is, the *Materials & Methods* section of a scientific paper in traditional format, or the *findings*, that is, the *Results* section (Malmsfors et al. 2004,

Gustavii 2008, Booth 1993). Starting with *what was done* has some merits, since it is the only aspect that is certain (e.g., there may be uncertainty about what to include in an *Introduction* or *Discussion*, but provided good records have been kept, there should be little doubt about *what was done*). Similarly, the researchers will have clear knowledge about their results (although their *implications* may be disputed). Furthermore, having written either of these sections, the resulting text can provide a framework on which to base the rest of the paper, to ensure that all sections are consistent.

However, this raises two problems. First, it is essential to know exactly what to include in these sections, which can only be decided after delimiting the study. Second, it is essential to describe the *Materials & Methods* (and *Results*) in a logical order. Generally, the optimal order is chronological, for reasons discussed later (although other approaches to organizing material *can* be used, see Matthews and Matthews 2008). However, investigators might only remember to analyze certain control samples that should have been analyzed in early stages of an investigation toward its end. In such cases, they would seem foolish if they presented what they did in the *true* chronological order, stating at the end of the *Materials & Methods* section *We then analyzed the controls, which we had previously forgotten to do.* Instead, it would be far better to state at an earlier point that *Both the extracts and controls were analyzed.* Similarly, a substance that has taken months to purify may be dropped, scraped off the floor, re-purified and then analyzed. In such cases we would not recommend stating *The substance was dropped, scraped off the floor, re-purified and then analyzed.* Instead, we would write, simply *The purified substance was then analyzed.* Thus, the *Materials & Methods* section should present *what was done*, or rather what would have been done if everything had been done correctly the first time in an ideal order, which may not coincide completely with the order in which everything mentioned was actually done.

For these reasons, a framework (which should be clear, simple and consistent) is required *before* starting to write this or any other section. Fortunately, such a framework can be constructed, for any study, by briefly describing the *rationale*, *objective(s)*, *what was done*, the *findings* and the *implications* of the study. The way in which such a framework can be used to compose each section of a paper is described in detail in following parts of this guide, but before doing so we should define these terms, recognizing that a scientific investigation is rarely a smooth progression from an initial rationale, through formulation of a set of testable hypotheses, to experiments that have been perfectly designed and executed, yielding perfectly analyzed and interpreted results. Thus, here:

- *Rationale* refers to the context or background of the study, as understood at the time of writing, which may not fully coincide with the initial rationale. For instance, the initial rationale may have been partly based on a misunderstanding of a previous author's work. If so, we would not recommend writing that certain *hypotheses were tested because we misunderstood Smith's conclusions*, but instead adjust the rationale. Note, this is quite different from cases in which a well-grounded hypothesis formulated from a sound initial rationale was tested and found to be false, for which there is no need to adjust the rationale.

- *Objectives* refers to the specific goals of the study as understood (with hindsight) at the time of writing, which again may not fully coincide with the initial objectives, since the goals may shift during the course of the study; some may be added, some may be dropped and others may change. Thus, for instance, it would usually be pointless for an author to describe the context of a hypothesis that he/she initially planned to test, but did not because there was insufficient time, except perhaps in the conclusion, if possible future analyses are mentioned.
- *What was done* refers to the experiments and analyses that were performed, in the order that they should ideally have been performed.
- *Findings* refers to the results from those experiments and analyses, and the conclusions that can be directly drawn from them.
- *Implications* refers to conclusions that can be indirectly drawn or inferred from the *findings*, for example, whether a tested substance could be viably used in a proposed application, with referenced comparisons to previously published findings.

Initially, statements describing these aspects can be very short. Indeed, writing short statements describing each of the aspects is essential for composing key sections of a paper (especially the *Title*, *Abstract* and *Conclusion*). In addition, a fuller, much more detailed framework can be very useful for checking that all aspects of the paper are consistent, in other words that:

- the *rationale* provides sufficient context to justify everything that was done
- the description of *what was done* details all the materials used, treatments applied and experiments for which results will be mentioned
- results of all experiments mentioned in the *what was done* section are covered
- all of the main *findings* are discussed and
- appropriate references have been added at appropriate places.

How such a framework can be drafted and applied in practice are the main concerns of the rest of this guide.

References

Booth V (1993) Communicating in science: writing a scientific paper and speaking at scientific meetings. Cambridge University Press, Cambridge, UK

Gustavii B (2008) How to write and illustrate a scientific paper. Cambridge University Press, New York, NY

Malmsfors B, Garnsworthy P, Grossman M (2004) Writing and presenting scientific papers. Nottingham University Press, Nottingham

Matthews JR, Matthews RW (2008) Successful scientific writing: a step-by-step guide for the biological and medical sciences. Cambridge University Press, New York, NY

Chapter 3
Drafting Papers

Writing a paper can seem an overwhelming task initially, since there are many points to include, and it may seem very difficult to arrange the information coherently and appropriately. However, scientific papers follow a standard format, with minor variations, based on discrete sections (*Abstract/Summary*, *Introduction*, etc.), in which information on various aspects of the study should be presented in a strict order. This format provides a highly convenient template, allowing *any* paper to be composed by writing brief, simple statements about the *rationale*, *objectives*, *what was done*, the *findings* and the *implications* of the study to be reported. In conjunction with sub-headings, if necessary, these statements can provide a framework showing precisely the information (and references) that needs to be included, and the order in which it should be presented, in each section. This chapter details how this systematic approach can be used to draft a paper describing any relatively simple study. How this can be done for complex studies is described in Chapter 4.

3.1 Arranging the Information

In addition to being clear and consistent, the information must be presented in any paper in a logical order. Fortunately, a standard format has been developed for presenting scientific studies with the following: a *Title*, *Abstract* and *Introduction* followed by *Materials & Methods*, *Results*, *Discussion*, *Conclusion(s)* and *References* sections, with varying numbers of *Figures* and *Tables*. Of course, there are many variations of the format. For example, the *Materials & Methods* section may be called the *Experimental* section, it may be placed after the *Discussion* and there may not be a formal *Conclusion(s)* section. However, nearly all science journals require some variant of this format to be followed. Even *Short communications*, in which most or all of these sections are merged into a single section, can generally be divided into such sub-sections, even if they are not formally separated by sub-headings.

This provides a valuable template for arranging information to maximize chances of publication and, as mentioned earlier, the key points that are going to be covered in a paper about a study can be easily encapsulated by briefly writing the following:

J. Blackwell, J. Martin, *A Scientific Approach to Scientific Writing*,
DOI 10.1007/978-1-4419-9788-3_3, © Springer Science+Business Media, LLC 2011

- *Rationale*: Detailed in the *Introduction*, often recapped at the beginning of the *Discussion*
- *Objectives*: Summarized at the end of the *Introduction*
- *What was done*: Summarized at the end of the *Introduction*, detailed in the *Materials & Methods* section
- *Findings*: Detailed in the *Results*, mentioned in the *Discussion*
- *Implications*: Detailed in the *Discussion*.

Brief statements of these aspects can also be used, with appropriate modification, to construct sub-headings for sections of the papers, which can be very useful for maintaining narrative flow (telling a coherent story), even in cases where sub-headings are not actually included in the text, as illustrated in Section 3.3.1.

Let us first consider how such brief statements can be used to construct a *Title* and *Abstract* (although in practice these components are often written last, partly because when a paper is written important aspects of the study that were initially forgotten may be remembered, and partly because they are the most critical parts of a paper).

3.2 The *Title* and *Abstract*

Providing all the key information is most difficult, but most essential, in the *Title* and *Abstract*, for the following reasons:

- they are the only parts of the paper that are read by many readers, and often the only parts that are freely available
- they have to summarize the study and be fully understandable without the rest of the paper
- they must be short
- they must show that the study has novel aspects.

Let us consider first how brief statements can help us to write a *Title* and *Abstract* for a simple hypothetical case study. Please note, this and all of the other hypothetical case studies are fictitious; they do not necessarily reflect real situations in any way. In addition, the examples are largely biological or ecological, partly because our core competence is in these fields (although we work closely with editors that are experts in other fields), and partly because biological and ecological examples are easier to understand for scientists generally than (for instance) quantum mechanical modeling. However, the procedures described to construct sections of papers can be applied to studies in virtually any scientific discipline. Points at which references should be included are noted, but specific references are not generally cited, since the studies are hypothetical. In addition, we have avoided use of complex technical terms, where possible, to ensure that the points raised can be understood by as many readers as possible.

3.2.1 Hypothetical Case Study 1

- *Rationale*: Substantial proportions of crops in northern Sucrosia are lost to grazing by deer (*Cervus unreal*) roaming from neighboring hills. Current methods for deterring the deer are criticized for being inhumane, expensive and/or ineffective. It has been postulated that applications of synthesized volatiles from urine of wolf (*Canis lupis* subspecies *imaginary*) could be cheaper and more effective than current methods, but first it is necessary to check that wolf urine is an effective deterrent.
- *Objective*: To test the hypothesis that wolf (*Canis lupis* subspecies *imaginary*) urine is an effective deterrent.
- *What was done*: Duplicate sets of three enclosures containing grass meadow surrounding plots of wheat, maize and pea crops were established. Wolf urine was applied around the perimeters of the plots in one set of enclosures, but not the other. Matched herds of deer were then introduced to each enclosure, and the proportions of the crops consumed in the plots surrounded by wolf urine and the other (control) plots were observed.
- *Findings*: The deer rapidly consumed the crops in the control plots, but completely avoided the plots that had been surrounded by wolf urine.
- *Implications*: Urine of wolf subspecies *imaginary* is an effective, odor-based deer deterrent.

Such a straightforward study, in which a single hypothesis was postulated and tested, and the results were very clear, would almost certainly be presented as a *Short communication*. However, as previously mentioned, even communications in which most or all of the text is merged into a single section can generally be divided into standard sub-sections (*Title, Abstract* etc.) even if they are not formally separated by sub-headings. Thus, let us consider first how these brief statements can help when writing a *Title*.

In such cases, one often sees titles such as:

> Effects of wolf urine on deer grazing crops

However, merely jotting down and slightly modifying the main *findings* or *implications* gives a much more informative title:

> Urine of wolf subspecies *imaginary* deters *Cervus unreal* deer from grazing crops in northern Sucrosia

Similarly, for instance:

> Ice on pavements increases risks of falls

is better than

> Effects of ice on pavements on frequencies of falls

Such titles tell the editor and referees of a journal immediately whether or not the paper is likely to be of interest to the readers, and thus if it is worth considering. However, it is not always possible to summarize *findings* or *implications* in a title.

For example, effects of diverse climatic, soil and other environmental factors on wheat yields in Sucrosia may have been examined. In such cases, it is often possible to write a clear, informative title by simply jotting down a modified form of the *objectives*, perhaps as follows:

> Multi-factorial analysis of the effects of environmental variables on wheat yields in Sucrosia

Let us now turn our attention to *Abstracts*. Providing all the critical information in the *Abstract* can often seem difficult, since many points *could* usually be included, but word limits are generally very tight.

Fortunately, jotting down a summary of the *rationale, objectives, what was done, findings* and their *implications* is also helpful when writing *Abstracts*. Indeed, if one or more of these aspects are not included, the resulting *Abstract* will be uninformative. In extreme cases one sometimes sees poor abstracts such as the following:

> We have examined effects of wolf urine on distributions of deer grazing. The results show that it strongly influences the plants deer choose to graze. We discuss the implications of these findings.

This does not mention the location of the trials, the treatment(s), or the *implications* that urine (and presumably volatiles in it) could provide effective crop protection from grazing deer. Failing to mention these points will seriously damage chances of the paper being accepted. The results *could* be interesting, but it is impossible to tell from this *Abstract*. Thus, the whole paper must be read to find out if it is worth reading, but many readers (including potentially the editor or referees of the target journal) will not bother.

The *Abstract* could be greatly improved, simply by mentioning (in order) the *rationale*, the *objective(s)*, *what was done*, the main *findings* and their *implications* (with a little modification), as follows:

First the *rationale*...

> Substantial proportions of crops in northern Sucrosia are lost to grazing by deer (*Cervus unreal*) roaming from neighboring hills. Current methods for deterring the deer have been criticized for being inhumane, expensive and/or ineffective. It has been postulated that applications of synthesized volatiles from urine of wolf (subspecies *imaginary*) could be cheaper and more effective than current methods, but first it is necessary to determine whether wolf urine is an effective deterrent.

Then the *objectives* and *what was done*...

> To test the hypothesis that wolf (subspecies *imaginary*) urine is an effective deterrent, duplicate sets of three enclosures containing grass meadow surrounding plots of wheat, maize and pea crops were established. Wolf urine was applied around the perimeters of the plots in one set of enclosures, but not the other. Matched herds of deer were then introduced to each enclosure, and the proportions of the crops consumed in the plots surrounded by wolf urine and the other (control) plots were observed.

Then the *findings*...

> The deer rapidly consumed the crops in the control plots, but completely avoided plots that wolf urine had been applied around.

And finally the *implications*..

> The results indicate that urine of wolf subspecies *imaginary* is an effective deer deterrent.

The resulting *Abstract* is already much more informative. All that remains to be done is to link the statements, and add a little detail, as follows:

> Substantial proportions of crops in northern Sucrosia are lost to grazing by deer (*Cervus imaginary*) roaming from neighboring hills. Current methods for deterring the deer have been criticized for being inhumane, expensive and/or ineffective. It has been postulated that applications of synthesized volatiles from wolf (*Canis lupis* subspecies *imaginary*) urine could be cheaper and more effective than current methods, since the deer are believed to have a strong aversion to sites marked by the wolf. However, first it is necessary to determine whether wolf urine really is an effective deterrent. Therefore, duplicate sets of three enclosures containing grass meadow surrounding plots of wheat, maize and pea crops were established. Wolf urine was applied around the perimeters of the plots in one set of enclosures, but not the other. Matched herds of deer were then introduced to each enclosure, and the proportions of the crops consumed in the plots surrounded by wolf urine and the control plots were observed. The deer rapidly consumed the crops in the control plots, but completely avoided the treated plots for two days. The results indicate that volatiles in wolf urine are effective deer deterrents, and if they are identified and synthesized they may provide cheap, potent crop protection from roaming deer.

It should be stressed that this study is purely illustrative, it does not necessarily reflect interactions in any real area.

3.2.1.1 Novel Aspects

As previously mentioned, a paper should have at least one novel aspect, and the novel aspect(s) should be highlighted, if possible, in both the *Title* and *Abstract*. For example, the finding that applying wolf urine could prevent deer grazing in Sucrosia would be the novel aspect in case study 1.

If you discover a completely new phenomenon or develop a new technique (e.g., time travel), you will not have to think much about the novel aspects. However, in other cases it is very important, because if there are no novel aspects the paper is unlikely to be accepted. Therefore, one may have to think carefully about any novel aspects of a study that can be highlighted.

For example, if you detect a compound in an annual plant that has only been found in trees previously, you could state as follows:

> This is the first time that this compound has been detected in a non-woody species, raising interesting questions about its physiological role.

Note, it is not necessarily essential to have data regarding its physiological role (although if you do have any they should be stated); merely mentioning that the compound has not been detected in non-woody plants previously, and it *may* have unsuspected functions, will greatly add to the potential interest.

Or, if a soil process has been explored in numerous ecosystems, but your group is the first to examine it in arid grasslands, you could state that

> Our study extends knowledge regarding the process to arid grasslands, and confirms that it has marked similarities across a wide spectrum of ecosystems.

or, alternatively,

> The results indicate that intriguing variations in the process are linked to variations in soil, climate or vegetation that warrant further investigation.

(This could also be used in a grant application).

The construction of a *Title* and *Abstract* from brief summaries of an additional, and somewhat more complex, case study is illustrated in the following text, and their construction for papers describing much more complex studies is illustrated in Chapter 4.

3.2.2 Hypothetical Case Study 2

The second hypothetical case study can be summarized as follows:

- *Rationale*: A previously minor insect pest (the weevil *Hylobius fabricated*) of young pine trees has begun to attack, aggressively, some (but not all) populations of *Pinus fictitious*, a threatened species of pine, in hilly parts of northern Sucrosia. Damage by pests is generally influenced by both environmental and genetic factors, thus there is a strong likelihood that both of these kinds of factors affect levels of damage in these populations. Furthermore, the invasive weed fictional knotweed (*Fallopia fictional*) has spread extensively, and may be strongly affecting interactions between plants and insects, in the region.
- *Objective(s)*: To determine factors responsible for variations in damage to fictitious pines caused by *Hylobius fabricated*.
- *What was done*: Correlations between levels of damage and both environmental and genetic factors were examined within and among three lightly damaged and three heavily damaged populations of the pine.
- *Findings*: Both environmental factors (especially coverage of the small invasive weed fictional knotweed, *Fallopia fictional*, and stand density) and genetic factors (alleles at two fictitious quantitative trait loci, i.e., stretches of DNA associated with certain traits, designated *res2* and *res4*) strongly and interactively affect levels of damage by the pest.
- *Implications*: Efforts to conserve the threatened pine species should focus on removing fictional knotweed and thinning stands to foster environmental conditions that do not promote attacks, and re-planting with saplings that carry *res2* and *res4* where necessary.

As illustrated for the previous hypothetical case study, the *findings* or *implications* can be used to construct a *Title*, as follows:

> Coverage of fictional knotweed, stand density and alleles *res2* and *res4* strongly affect damage in *Pinus fictitious* populations by *Hylobius fabricated*.

Similarly, too, an *Abstract* can be composed from modified forms of the brief statements regarding the *rationale*, the *objective*, *what was done*, the *findings* and their *implications*, sequentially, as follows:

The *rationale*...

Hylobius fabricated, a previously minor pest, has recently begun to attack aggressively some, but not all, populations of the threatened pine species *fictitious* in Sucrosia.

The *objectives*...

To identify factors responsible for variations in damage caused by this insect, we examined correlations between levels of damage and both environmental and genetic factors.

The *findings*...

The results show that both types of factors (most importantly, coverage of the invasive weed *Fallopia fictional*, stand density and alleles *res2* and *res4*), and interactions between them, strongly affect levels of damage.

The *implications*...

The findings indicate that efforts to conserve this threatened pine species should include removing *Fallopia fictional* and thinning to foster environmental conditions that do not promote attacks, and re-planting where necessary with saplings carrying alleles *res2* and *res4*.

Having done this, an informative *Abstract* can be composed by linking the statements, and adding appropriate details from notes compiled during the study, as follows:

Hylobius fabricated, a previously minor pest, has recently begun to attack aggressively some, but not all, populations of threatened fictitious pines in Sucrosia. To identify factors responsible for variations in damage caused by this insect, we examined correlations between levels of damage and both environmental and genetic factors in three lightly attacked and three heavily attacked populations. The results show that both types of factors, and interactions between them, strongly affect levels of damage. Most importantly, Analysis of Variance indicates that coverage of the invasive weed *Fallopia fictional*, stand density and the presence of alleles *res2* and *res4* accounted for ww, xx, yy and zz% of the variance in observed damage levels, respectively. The findings indicate that efforts to conserve this threatened pine species should include removing *Fallopia fictional* and thinning to foster environmental conditions that discourage attacks by the pest, and replacing trees killed by the pest with saplings carrying alleles *res2* and *res4*. However, further research is required to identify the reasons why the identified factors influence levels of damage and optimal ways to clear the weed.

3.3 The *Introduction*

If a paper has an interesting *Title*, and an informative *Abstract* indicating that the paper provides novel information of interest to readers of the target journal, it will already be well on the way to being accepted. However, there are several other sections to consider, the next being the *Introduction*.

The *Introduction* should explain why the study was conducted (and why the subject is important) by describing: the context of the study, with references to previous studies; the phenomena investigated; the reasons for studying them; and (very briefly) how they were investigated.

This should be done by first describing the *rationale* of the study in much more detail than in the *Abstract*, and *then* briefly outlining the *objectives* and *what was done*. For this purpose, fuller statements about the *rationale* are required that can be used to construct convenient sub-headings (which may be subsequently deleted), as illustrated later for the two hypothetical case studies described thus far. There are also a number of potential pitfalls to avoid, some of which will be mentioned during this section on *Introductions*, which focuses on the case studies already mentioned, and some later, in the corresponding sections for more complex studies (Chapter 4).

3.3.1 Hypothetical Case Study 1

To refresh readers' memories (and avoid any need to go back to check), the brief *rationale* for this study was as follows:

> Substantial proportions of crops in northern Sucrosia are lost to grazing by deer (*Cervus unreal*) roaming from neighboring hills. Current methods for deterring the deer have been criticized for being inhumane, expensive and/or ineffective. It has been postulated that applications of volatiles from urine of wolf subspecies imaginary could be cheaper and more effective than current methods, but first it is necessary to check that wolf urine is an effective deterrent.

The *objective*: To test the hypothesis that wolf (subspecies *imaginary*) urine is an effective deterrent.

What was done: Duplicate sets of three enclosures containing grass meadow surrounding plots of wheat, maize and pea crops were established. Wolf urine was applied around the perimeters of the plots in one set of enclosures, but not the other. Matched herds of deer were then introduced to each enclosure, and the proportions of the crops consumed in the plots surrounded by wolf urine and the control plots were observed.

As outlined earlier, the *Introduction* for this (and any other) study should start by expanding the *rationale*. Therefore, the next step is to jot down very brief sub-headings for each aspect of the *rationale* that needs to be covered. For instance:

- *Grazing deer cause substantial losses of crops in Sucrosia*
- *Current deterrents are inhumane, expensive or ineffective*
- *Wolf urine volatiles could be better deterrents*
- *The need to check that wolf urine is an effective deterrent.*

Following these sub-headings, an initial version of the *Introduction* (which will need to incorporate further details later) can be drafted by expanding the *rationale*, based on notes made during the study, as follows:

- *Grazing deer cause substantial losses of crops in Sucrosia*
  Grazing by deer (*Cervus imaginary*) roaming from neighboring hills sometimes causes substantial losses of wheat, maize, pea and other crops in northern Sucrosia (followed by illustrative data on the amounts and costs of the losses, with references).
- *Current deterrents are inhumane, expensive or ineffective*
  Farmers attempt to deter the deer using fences, culls and various kinds of traps, but these methods have been criticized for being inhumane, expensive and/or ineffective (followed by illustrative data, with references).
- *Wolf urine volatiles could be better deterrents*
  However, Deerman (2004, fictitious essential reference) has postulated that applications of volatiles from wolf (subspecies *imaginary*) urine could be cheaper and more effective than current methods, since these wolves are the major predators of the deer and the deer are believed to exhibit strong aversion to sites marked by the wolves. In addition, various studies have shown that herbivores avoid areas that have been recently marked with predator urine, although there are wide variations in their responses (followed by illustrative examples from the literature).

  Two major kinds of repellents in predator urine have been identified: taste repellents and odor repellents. The former only repel herbivores after they have started eating food that has come into contact with them, while the latter may repel from a distance (followed by further illustrative examples from the literature). Clearly, spraying crop plants with either wolf urine or taste repellents derived from it is not an attractive option. Therefore, any compounds isolated from the urine to be used in this way would have to be volatiles that could be applied around crops rather than onto them. However, before seeking active wolf urine volatiles (if present) it is necessary to check that wolf urine really is an effective deterrent.

Presenting the *rationale* in this manner clearly outlines the context of the study and highlights its importance. It should be noted that this was a very simple study in which a single clear hypothesis was formulated, very straightforward methods were applied and very clear results were obtained. Thus, there should not be any need to mention pros and cons of the methodology for this case study. In other cases discussing such aspects is *critical*, for reasons that are discussed in later sections.

The draft of the *Introduction* should conclude with a *brief* description of the *objectives* and *what was done*:

> To test the hypothesis that wolf (subspecies *imaginary*) urine is an effective deterrent, it was applied around the perimeters of one set of plots containing wheat, maize and pea crops in enclosures, but not around perimeters of plots in duplicate enclosures. Deer were then introduced to each enclosure, and the proportions of the crops consumed in the plots surrounded by wolf urine and the control plots were recorded.

A draft of an *Introduction* such as this is not complete, since details and references need to be added. However, it is very useful because it pinpoints the data and references (essential and illustrative) that need to be added, and where they should be placed. The following parts of the guide show how to construct initial drafts of *Introductions* for three other case studies, and other sections of papers for this and the other case studies, to illustrate various general considerations.

3.3.2 Hypothetical Case Study 2

As described earlier, the *rationale* of this study was as follows:

> A previously minor insect pest (*Hylobius fabricated*) of pine trees has begun to attack, aggressively, some (but not all) populations of *Pinus fictitious*, a threatened species of pine, in Sucrosia. Levels of damage by pests are known to be influenced by both environmental and genetic factors, so there is a strong likelihood that both of these kinds of factors affect levels of damage in these populations. Furthermore, the invasive weed fictional knotweed (*Fallopia fictional*) has spread extensively, and may be strongly affecting interactions between plants and insects, in the region.

The *objective*: To determine factors responsible for variations in damage caused by *Hylobius fabricated*, which has recently begun, aggressively, to attack fictitious pines in Sucrosia.

What was done: Correlations between levels of damage and both environmental and genetic factors were examined in three lightly damaged and three heavily damaged populations of the pine.

A kind of construction that should generally be avoided when writing an *Introduction* is to mention *what was done* in each part of the study too soon. For instance, in this case, the *Introduction could* begin as follows:

> A previously minor insect pest (*Hylobius fabricated*) of pine trees has begun to attack, aggressively, some (but not all) populations of *Pinus fictitious*, a threatened species of pine, in Sucrosia. Levels of damage by pests are known to be influenced by both environmental and genetic factors, so there is a strong likelihood that both of these kinds of factors influence levels of damage in affected populations. Therefore, in the presented study both environmental and genetic variables that affect levels of damage by the pest were examined in six populations of the pine. The environmental factors were factors known to be associated with susceptibility to various insect herbivores in several related pine species...

Although this kind of construction is often used, it introduces repeated jumps from general considerations (e.g., that both environmental and genetic variables influence levels of damage in plant–insect pest interactions) to *what was done* (e.g., that both kinds of variables were examined in the selected populations) and back to general considerations. Such jumps between the *rationale* (context) of the study and the specific *objectives* disrupt the narrative flow (for more discussion of narrative flow, see Section 5.5), making the paper unnecessarily repetitive and difficult to read. Sometimes this problem is compounded by authors outlining some of the context of a study in the *Introduction* and then describing the *objectives* for the corresponding part of the study:

> The objectives of the present study were to examine factors that may affect the level of damage caused by this pest.

Then perhaps discussing environmental factors that affect resistance, before stating:

> Therefore the objectives of this study were to assess effects of environmental factors that are known to influence frequencies and levels of attacks by insects on other pines.

Then discussing genetic factors that affect resistance, before stating:

> Thus the objectives of this study were to assess effects of genetic factors that are known to influence attacks by insects on other pines.

Finally, they may repeat, at the end of the *Introduction*:

> The objectives of the present study were to assess the effects of environmental factors on levels of attack by *Hylobius fabricated* on fictitious pines in Sucrosia. In addition, we tested correlations between levels of damage caused by the insect and genetic factors that are known to affect interactions between pests and several other pine species. For this purpose we analyzed samples of pines in six populations in Sucrosia, and measured environmental variables at each of their sites.

The narrative flow is severely disrupted in such *Introductions* because passages setting the context of the study are separated by sentences repetitively describing the *objectives*. Hence it is difficult to link context-setting passages, making the text disjointed and irritating editors, referees and readers. Furthermore, stating *objectives* generally indicates that the *Introduction* is ending. Therefore, it is surprising when it continues with further text describing more of the context, and when the authors state a second set of *objectives*, readers may think, *This is strange, this is not what they said the objectives were in the previous paragraph.* If a third or fourth set of *objectives* is subsequently stated, they may think, *I wish they would make up their minds what their objectives were; if they can't decide, how do they expect me to work out what they were doing?*

To improve the narrative flow it is generally much better to start by fully describing the context of the study, starting with the most general aspects and then more specific contextual aspects, *before* outlining the specific *objectives* of the study and *what was done*. Fortunately, as illustrated for case study 1, the context can be clearly and concisely expressed (and the narrative flow can be maintained) in the *Introduction* of any paper, by using simple statements regarding the *rationale*, the *objectives* and *what was done* sequentially, with the assistance of sub-headings (and flow charts for complex studies, see Figs. 4.1 and 4.2).

Thus, the drafting of an *Introduction* for case study 2 should begin by jotting down sub-headings regarding the *rationale*:

- Hylobius fabricated *has begun to cause severe damage to some populations of fictitious pine in Sucrosia*
- *Environmental factors associated with damage to pines by pests*
- *Genetic factors associated with damage to pines by pests*
- *Pros and cons of the methodology used.*

Accordingly, an *Introduction* for case study 2 should begin by expanding the *rationale*, using relevant details, by stating that a variant of the previously minor pest *Hylobius fabricated* has begun to cause severe damage to some populations of fictitious pine in Sucrosia, and note that this is a serious concern because the pine is a threatened species, citing essential references, to establish the significance of the study.

It should then outline the reasons for focusing on the factors considered, stating first that levels of damage by pests are known to be influenced by environmental factors, providing illustrative examples from the literature.

In addition, since frequencies of alleles at homologous loci known to be associated with resistance to insect pests in related tree species were examined, details regarding these factors, the associated traits (e.g., bark thickness, the size and distribution of resin ducts, and the abundance of resin produced in response to insect pests) and the interactions with pests that they affect in the related species should also be mentioned, again with illustrative references.

In contrast to the *Introduction* for the previous very simple case study, the pros and cons of methods that could be used to assess the examined variables should also be discussed, and the advantages of the methods used should be highlighted, if objections regarding their use could be raised. Notably, methods used to assess levels of damage to trees should be mentioned. This is often done by ranking damage to individual trees in classes such as percentage intervals of foliage losses in a visual survey (although aerial surveillance methods are now being increasingly used), then constructing frequency distributions of numbers of trees in each damage class. This is relatively quick and convenient and can provide robust data, but it has the obvious disadvantage of being subjective. Thus, these advantages and disadvantages should be outlined, together with evidence (and illustrative references) that the methods you used are widely applied, to justify your use of them. However, it is not necessary (or even desirable) to state at this point that you did use any particular methods, since your methodology will be described in detail later.

Alternatively, the choice of methods could be justified in the *Materials & Methods* or *Discussion* sections. The *Materials & Methods* section may be the most suitable location if the reasons for using a selected method can be very briefly stated, for example, *Damage to trees was ranked following the procedure and criteria described by Vasquez* (fictitious reference), *which has been shown to provide robust classifications* (further essential references), while the *Discussion* may be optimal if the choice of methodology may have affected the results. The latter may be particularly helpful to your case if other choices would have heightened observed correlations, for example, *It should be noted that damage to trees was ranked following the procedure and criteria described by Vasquez* (fictitious reference), *which has been shown to underestimate damage compared to other possible approaches* (other essential references). *Hence, the observed correlations between damage and both environmental and genetic factors would probably have been enhanced if other methods for quantifying damage had been used.* Otherwise, it is probably best, generally, to justify methodological choices at an early stage, in the *Introduction*. Of course, the preferences of the target journal in this respect should also be considered, as always.

Finally, the *objectives* and *what was done* should be briefly summarized, outlining both the approach adopted to explore factors responsible for the variations and the reasons for focusing on the investigated populations. As previously mentioned, these reasons should be rational. It may be, for instance, that the *real* reason they were chosen is that they were the closest populations to your laboratory. We would not state that this was the reason for choosing them. Instead, we would state, for instance:

> Three severely attacked populations, and three lightly attacked control populations in a restricted area, where there was limited variation in climate and soils, were selected to examine environmental and genetic variables that influence the severity of attack by the pest (excluding climate and soil, since these variables cannot be readily altered in natural ecosystems).

A pitfall to avoid is omitting crucial information, which frequently happens because authors are too familiar with the subject. For instance, we have frequently encountered examples such as the following:

- papers about monitoring compounds, for example, polycylic aromatic hydrocarbons (PAHs), that explain in detail the methods used, but fail to describe the sampling sites, or that PAHs can cause serious health problems (and thus why the study is important)
- papers about water use efficiency by crop plants that do not mention the location of the study (although water use efficiency is irrelevant in a swamp, but critical in a desert)
- papers about treatments that could make small percentage savings in costs of industrial processes that do not mention that the total savings could amount to millions of dollars.

Often in such cases the authors assume that readers, referees and the journal editor will already know about crucial aspects that have not been explicitly mentioned. However, readers may not realize their significance unless they are clearly stated. If so, the importance of the study will be missed, so the chances of the paper being accepted may be severely reduced. It should be noted that referees and editors of journals also sometimes fail to see the need to mention crucial aspects explicitly (which is one of many reasons that some journals have lower impact factors than other, otherwise comparable journals), and may insist that passages describing the *rationale* or *objectives* of the study be deleted. However, the risk of a study being rejected because its significance has not been clearly stated far outweighs the risk of being asked to delete a couple of paragraphs.

It is also highly preferable to highlight the full significance of the subject at the *start* of the *Introduction*, *not* half way through it or close to its end. For example, if the feasibility of constructing defenses to protect a coastal town (Somewhere-on-Sea) from the sea following anticipated rises in sea level due to global warming has been examined, opening remarks such as

> According to climate models, anticipated global warming is likely to raise sea levels by between xx and yy m in the next zz years (refs.). If so, Somewhere-on-Sea and many other towns will be severely flooded, therefore there is a clear need to examine the feasibility of improving sea defenses to protect these towns, in addition to taking steps to reduce global warming.

are generally better than

> In the study presented here we examined the feasibility of improving sea defenses for Somewhere-on-Sea by analyzing the topography of the coast in the surrounding region and considering structures that could prevent flooding...

and then describing the methodology that can be applied in such studies before stating, later:

> This is important because sea levels are expected to rise by between xx and yy m in the next zz years (refs.).

The first alternative immediately establishes the significance of the study, and the reasons for undertaking it, while the second adds the most important element of the *rationale* almost as an afterthought.

Having discussed these pitfalls, let us consider the *Materials & Methods* section.

3.4 The *Materials & Methods* Section

The *Materials & Methods* section of a paper must relate, clearly and in much more detail than in the final remarks of the *Introduction*, *what was done* in the study, providing sufficient information to allow others to repeat it. This is common knowledge, but as with *Introductions*, authors often omit crucial information, because they think it is obvious or they simply forget to include it. For instance:

- authors sometimes say they cultured organisms, but do not mention the medium and/or kind of containers used
- some state that plants were grown 'outside', but do not mention where, at what time of year, and/or under what conditions
- others say they measured pollutants by gas chromatography-mass spectrometry (GC-MS), but do not describe the equipment or settings used.

Such omissions create a bad impression, which may cause a referee or editor to simply reject the paper. However, although including all the required information clearly and concisely is not always straightforward, there are several ways to make it easier.

The first, as already mentioned, is to keep good records of *everything that was done*. Everyone knows this, but people often forget, and instead scribble down important points about treatments or analytical protocols on scraps of paper, or merely think that they will remember them and hence do not need to record them carefully. This is a great error, which often results in people spending days trying to find the scraps of paper or trying to remember precisely what they did. In worst-case scenarios, it can even lead to people having to repeat experiments in order to recover information. One should always keep samples as long as possible for the same reason. As an example, a colleague once completed a study and as the final step photographed plants used in the investigation, then threw them away. However, due to a technical problem the photographs had not been saved, so he had to grow another set of plants, and in the meantime another group working in another country published results very similar to those he had obtained.

When drafting the *Materials & Methods* section of a paper, the descriptions of the materials and the methods used should be set out methodically (preferably chronologically, or in the chronological order that they *should* have been used) to help ensure that all of the important points are mentioned. If information is presented non-chronologically, it can be very confusing. For instance, authors sometimes describe plants they used in an experiment, then say that they subjected *the extracts* to preparative High Performance Liquid Chromatography (HPLC), before describing how the plants were extracted. The response of the reader is then *What extracts?* since none have been mentioned as yet. This also creates a bad impression that may increase the likelihood that the paper will be rejected. Further, it can be simply avoided by describing, sequentially, how the plants were grown or collected from the field, the treatments applied, how they were sampled and samples were stored (if not used immediately), the extraction techniques and finally the methods used to analyze the samples.

Of course, this may not be straightforward if the procedures were complex, but if standard techniques were employed, previous papers by other authors can be used as templates. For example, if polymerase chain reaction (PCR) amplifications were used, one can simply change the names of the sequences, primers, temperatures and so on in the corresponding section of a previous paper. Similarly, if HPLC or GC-MS was applied, one can simply use descriptions of the protocols applied in another study, changing the names of the equipment used and the settings as appropriate.

It should be noted that this is quite different from copying passages of a previous paper's *Introduction* or *Discussion*, which should not be done (or at least only done carefully, introducing sufficient variations in phrasing and the order in which information is presented, giving due acknowledgements when citing other authors' conclusions, to avoid accusations of plagiarism; for more details see Section 5.6). This is because there are only a few ways of coherently describing protocols for standard techniques, so all of them have been repeated with minor variations many times, hence plagiarism is not an issue when describing standard protocols.

In addition, since it is essential to provide sufficient information for the experiments to be repeated, the suppliers (and type or grade) of any reagents and equipment used should be specified, and if biological materials have been used, their sources should be mentioned. For example, if you acquired certain mutants from a library, it should be cited, and if you acquired samples from the field, the way the material was identified (e.g., by morphological examination of the organisms or genetic analysis) needs to be stated. Similarly, the developmental stage of the sampled material and/or organs or tissues should be specified. For example, authors often state that they sampled leaves of plants or insect larvae to measure contents of compounds of interest in them (perhaps in a metabolomic investigation) but do not state the developmental stage of the leaves or larvae. As with other omissions, this creates a poor impression and reduces the chances of papers being accepted.

The *Materials & Methods* section of a paper can be drafted using the brief statements of *what was done* with additional details to obtain sub-headings (which may be deleted later), in chronological sequence. This process is illustrated later, using the two hypothetical studies mentioned thus far as examples.

3.4.1 Hypothetical Case Study 1

The brief summary of *what was done* in this study was as follows:

> Duplicate sets of three enclosures containing grass meadow surrounding plots of wheat, maize and pea crops were established. Wolf urine was applied around the perimeters of the plots in one set of enclosures, but not the other. Matched herds of deer were then introduced to each enclosure, and the proportions of the crops consumed in the plots surrounded by wolf urine and the control plots were observed.

Sub-headings, based on fuller notes, may be as follows:

- *Collection and fractionation of wolf urine*
- *Establishment of enclosures and test crop plots*
- *Trials of the urine's deterrent effects.*

Thus, following these sub-headings, the *Materials & Methods* section for this study should sequentially describe the following

- the methods used to collect the wolf urine and the number, gender and age of the wolves it was collected from
- the size of the enclosures used to test the deterrent effects of the urine, and their meadow plant communities (with references to show that they were typical for the region), the size of the crop plots within them, and the developmental stage of the crops, with brief details of their cultivation
- the way the urine was applied around the perimeters of plots in one set of enclosures; and the age, gender, numbers and husbandry of the herds of deer used in the trials.

3.4.2 Hypothetical Case Study 2

Here, *Correlations between levels of damage and both environmental and genetic factors were examined within and among three lightly damaged and three heavily damaged populations of fictitious pine.*

Sub-headings could be composed as follows

- *Sites of selected populations*
- *Sampling sites within the areas containing the selected populations*
- *Variables and methods used to assess the status of the populations*
- *Variables and methods used to assess environmental factors*
- *Variables and methods used to assess genetic factors*
- *Statistical tests used to examine correlations between measured variables.*

Following these sub-headings, the *Materials & Methods* section should describe, sequentially

- The sites of the selected populations, in terms of geographical location (latitude, longitude and topology) and climate (providing references for the data, and the years they cover, if you did not take the measurements). Sources of any further data acquired from previously published sources (e.g., regarding soil characteristics) should also be mentioned.
- Sampling sites in the areas covered by the populations, criteria used to select them and the numbers and layout of sampling plots within them.
- Methods used to assess the status of the fictitious pines (e.g., stand density, size and age distributions and levels of damage by the pest within them) and levels of damage within them.
- Methods used to assess other environmental variables (e.g., soil variables – if you have not relied on previously published data – the plant community composition, cover of the species in the canopy and other vegetation layers, density of the ground layer, litter thickness and frequencies of fallen trees).
- Methods used to sample pines in the populations to collect genetic material, the sequences used for genetic analysis, the databases used to identify homologous sequences associated with insect resistance in related tree species and any methods used to estimate allele frequencies and genetic diversity of the populations, with essential references.
- Finally, the statistical methods used to examine correlations between levels of damage within and among the populations and both the environmental and genetic factors. If the genetic diversity of the populations has also been evaluated, to assess the possibility that it influences attack levels, the methods used to do so should also be clearly described, with essential references.

Having written a *Materials & Methods* section (after adding appropriate details and references), it should ideally be read by an expert in the subject (e.g., water use efficiency of plants in arid environments), to ensure that it is comprehensible and nothing important has been missed. In addition, it should be read from the perspective of a reader with an interest in the general subject (e.g., a plant physiologist), but not expertise in the specific subject. Better still, a real person with general interest should be asked to read it. Such an individual cannot be expected to understand all of the technical details, but if his or her feedback is *I have no idea how or why you did this*, it does not mean that the reader is stupid – it means that your paper needs to be modified, by simplifying the text, eliminating jargon if possible (see Section 5.1) and adding more information or references if necessary. Otherwise, there is a risk that your paper will only be understood by a few specialists, thus restricting the numbers of people who may read (and cite) it.

3.5 The *Results* Section

The *Results* section of a paper must describe the *findings*. This too may not be straightforward if multiple kinds of relationships and/or variations have to be described. However, as for other sections, the general principles of writing clearly and simply still apply, and the process can be facilitated by using brief statements describing the *findings* obtained in each stage of the investigation, in the (chronological) order in which the procedures used were presented in the *Materials & Methods* section as a framework. This will ensure that the results obtained in each stage of the investigation are presented in the same order the methodological steps were presented in the previous sections and avoid errors such as describing results of analyses of purified extracts before mentioning how the extracts were prepared (and thus irritating referees and editors).

As for *Materials & Methods* sections, it is also often possible to use published papers as templates. For instance, if dioxins in a Chinese ecosystem have been examined, parts of papers relating to studies of dioxins in the US, such as descriptions of GC-MS spectra, could be adapted with appropriate modifications. Therefore, we recommend keeping a list of papers that clearly describe techniques that you use, and results obtained with them, as templates to help when writing sections of papers describing standard methods and results obtained when using them. In addition, figures and tables should be *judicially* used to present results. For instance, including large quantities of information in the text that could be more conveniently summarized in tables or appropriate figures should be avoided. Thus, examples of such tables and figures in previous papers should also be kept, as templates. Similarly, very large tables may be better placed in *Appendices* or *Supplementary Information* available online than in the *Results* section.

The use of figures and tables is considered in further detail later (see Section 3.9), but before that let us consider how the *Materials & Methods* sections can be used to draft *Results* sections that follow a coherent order, for each of the two hypothetical case studies considered thus far.

3.5.1 Hypothetical Case Study 1

The *Results* section for the hypothetical study of the efficacy of wolf urine as a deterrent of deer grazing in Sucrosia would be extremely simple to write since it would have to describe only the single main finding: that the deer rapidly consumed untreated crops but, for 2 days, completely avoided crops in plots after wolf urine had been applied around the perimeters.

3.5.2 Hypothetical Case Study 2

For this study of factors affecting damage caused by *Hylobius fabricated* to fictitious pines, the *Results* section should describe the *findings* fully, following the order in which the procedures used to obtain them were described in the *Materials & Methods* section, and thus sequentially presenting

- measurements of the status of the fictitious pines (e.g., stand density, size and age distributions and levels of damage by the pest)
- measurements of environmental variables (e.g., the community composition, cover of species in the canopy and other vegetation layers and perhaps soil variables or frequencies of fallen trees) and any information used in the analyses obtained from other sources (with essential references)
- the frequencies of examined alleles, and any other calculated genetic parameters, such as within- and among-population diversity parameters, with confidence limits
- finally, the correlations between damage levels, and each of the measured environmental and genetic variables (and their interactions), together with indications of their statistical significance.

3.6 The *Discussion* and *Conclusion(s)*

According to standard text books, in the *Discussion* one should consider:

- What do the results mean?
- How do they relate to other published results?
- What are the implications?
- What problems occurred?
- What improvements could be made?
- What more needs to be done?

However, the *real* significance of the *Discussion* is that it is the last chance (apart from the *Conclusion(s)*, see later) to show that your study is important and convince the editor and referees of the target journal that the study should be published, by focusing on the *implications* of the results. This is in contrast to the *rationale* and *objectives* of the study (which should largely be covered in the *Introduction*), *what was done* (which should largely be covered in the *Materials & Methods* section) and the *findings* (which should largely be covered in the *Results* section), as previously mentioned. In a paper describing such a straightforward study as case study

1, the *Discussion* would probably be merged with the *Results*, possibly in a single paragraph, as follows:

> The deer completely avoided crops in plots that had been surrounded by wolf urine for two days, but rapidly consumed crops in the control plots. This is consistent with previous reports that prey species frequently avoid areas marked by predators' urine (followed by illustrative examples from the literature, with references). However, the effects of the urine in this case appear to be particularly potent (followed by further comparative examples from the literature). Furthermore, the effects appear to be due to volatile components in the urine, rather than taste repellents, since the deer did not enter the plots protected by the urine applications at all for two days (followed by further examples of similar cases from the literature of volatiles affecting the behavior of prey species, and contrasting examples in which they were repelled only after eating food that had come into contact with predator urine). This is highly encouraging, since it suggests that it should be possible to extract (and subsequently synthesize) volatiles from the urine that may be potent deterrents.

However, usually there is inevitably some overlap between the *implications*, *rationale* and *findings* in a *Discussion*, since it is impossible (for instance) to discuss the *implications* of the results without briefly recapping the major *findings*. It is also often sensible to remind readers of the general importance of the subject in a long paper (partly because they may have forgotten by the time they start reading the *Discussion*, and partly because it often helps the 'narrative flow') by very briefly recapitulating part of the background (*rationale*) of the study. For instance, a draft of the *Discussion* for case study 2 might start as follows:

> The previously minor pest *Hylobius fabricated* has begun to attack, heavily, some (but not all) populations of fictitious pines, a threatened species (essential reference), in Sucrosia. Therefore, there is an urgent need to identify factors that contribute to high levels of damage by this insect. In an attempt to discover some of the factors that may be involved, correlations between potentially contributory factors and levels of damage in six selected populations of fictitious pine in Sucrosia were examined.

The main *findings* of the study then need to be briefly recapitulated, since otherwise it is difficult to discuss their *implications*. However, for a complex study this is correspondingly more complex than for a simple study, since multiple *findings* must be recapitulated, and the *implications* of each finding must be considered. There are two main ways of doing this. If all of the results can be concisely described in a few sentences, they could be outlined first. For instance, for case study 2 the *Discussion* could continue as follows:

> As reported earlier, the results clearly show that both environmental and genetic factors influence the levels of damage.

However, *this should be avoided*, since much more detail needs to be added after the opening remarks in the *Discussion*, thus this information will have to be repeated in subsequent paragraphs, and again in any concluding remarks. Hence, whenever there is more than one main finding to discuss, it is generally far better to recap each of the main *findings* you wish to consider, separately, in the order they were presented in the *Results* section and discuss how they compare with results of previous studies, and their *implications*, before progressing to the next main finding.

Thus, for example, a draft of the *Discussion* for case study 2 could continue:

The results indicate that environmental factors influence levels of damage caused by the pest to fictitious pines in Sucrosia. Notably, damage appears to be heavy at sites where the invasive weed fictional knotweed is abundant and the stand density high (followed by data showing the strength of the correlations, within and among populations, and the variance explained). This is consistent with previous reports (followed by illustrative references regarding similar correlations in pine–pest interactions). Any other correlations detected between environmental variables and damage then need to be mentioned (followed by further illustrative references describing similar or contrasting correlations). Similarly, environmental variables that were not correlated with damage should be noted, and indications whether the lack of correlation is consistent, or conflicts, with previous findings. In addition, possible explanations for the observed correlations should be described (with support, if available, from relevant references).

A subsequent paragraph could state the following:

The results also show that genetic factors affect levels of damage. Notably, alleles *res2* and *res4*, which are strongly associated with traits that influence resistance to insect pests, including bark thickness, the magnitude and distribution of resin ducts and the abundance of resin produced in response to insect herbivores (followed by data showing the strength of the correlations, within and among populations, and the variance explained). This is also consistent with previous reports (followed by illustrative references regarding similar correlations affecting pine–pest interactions). Any other correlations detected between genetic variables and damage then need to be mentioned (followed by further illustrative references describing similar or contrasting correlations). Again, possible explanations for the observed correlations should also be described (with support, if available, from relevant references).

The *Discussion* of these, and all studies, should end with summarizing or concluding remarks, which may or may not be presented in a separate section entitled *Conclusion* or *Conclusions*. Otherwise the paper will be like a joke without a 'punch line'. The concluding remarks also provide the last chance of all to persuade doubtful referees or editors that the paper should be published. The ideal format will depend on whether or not the target journal requires a specific *Conclusion(s)* section and the information presented in the last section of the *Discussion*, especially if there is not a specific *Conclusion(s)* section, since narrative flow should then continue from the last sentences of the *Discussion*. However, the concluding remarks should usually recap, with some modifications, the *objectives*, *findings* and their major *implications*. For instance, the hypothetical studies addressed thus far in this guide could be concluded as follows, for case study 1:

The findings show that urine of the wolf *Canis lupus* subspecies *imaginary* is a powerful deterrent of grazing by deer (*Cervus imaginary*). Small applications of the urine along the perimeters of crop plots at 2 m spacing were sufficient to completely exclude deer in the trial, while they rapidly consumed the crops in the control plots. Full-scale trials are required to confirm these findings, but the results are highly promising.

and for case study 2:

The results indicate that both environmental and genetic factors influence levels of damage caused by *Hylobius fabricated* to fictitious pines in Sucrosia. Levels of damage appear to be heavy at sites where fictional knotweed is abundant and the stands are dense. In addition, the alleles *res2* and *res4* appear to confer substantial resistance to the pest, in accordance with observations by several authors that homologous genes confer resistance to various insect pests in related pine species. Thus, if these threatened pines are to be conserved in Sucrosia (and potentially other parts of the world), it is important to ensure that programs to manage remaining populations include clearing knotweed, thinning and planting saplings that carry *res4* and *res4* in re-planting programs.

3.6.1 Combined Results & Discussion Sections

As mentioned earlier, for a very straightforward study, such as case study 1, the *Results* and *Discussion* sections would probably be merged into a single section (possibly consisting of a single paragraph in a *Short Communication*). However, many journals also require results of larger, more complex studies to be presented and discussed in a combined *Results & Discussion* section, rather than separate sections. Thus, the way that such a section should be composed should also be considered. Generally, as for a *Discussion*, it is often wise to begin by briefly recapitulating the *rationale* of the study, to refresh readers' memories of the importance of the study. Thus, for case study 2 it could begin in the same way as a separate *Discussion* section as follows:

> The previously minor pest *Hylobius fabricated* has begun to attack, heavily, some (but not all) populations of fictitious pines, a threatened species (essential reference), in Sucrosia. Therefore, there is an urgent need to identify factors that contribute to high levels of damage by this insect. In an attempt to discover some of the factors that may be involved, correlations between potentially contributory factors and levels of damage in six selected populations of fictitious pine in Sucrosia were examined.

Having outlined the *rationale* in this manner, a temptation is to continue by approaching the section in the same manner as if writing separate sections, that is, by presenting all of the results first, and then discussing them. However, while this approach is perfectly acceptable (and indeed the standard procedure) if the sections are separate, it is not usually optimal for a combined section, since it can lead to unnecessary repetition, and make it unnecessarily difficult to read. Instead, it is usually better to present the results, and discussion concerning them, sequentially. For instance, for case study 2, the rest of a combined *Results & Discussion* section could present:

- Measurements of the status of the fictitious pine populations (e.g., stand density, size and age distributions and levels of damage by the pest), followed by details of the correlations between damage levels, and each of the measured variables (and their interactions), together with indications of their statistical significance.
- Measurements of environmental variables (e.g., the community composition, cover of species in the canopy and other vegetation layers and perhaps soil variables or frequencies of fallen trees) and any information used in the analyses obtained from other sources (with essential references). Again this should be followed by details of the correlations between damage levels and each of the measured variables (and their interactions), together with indications of their statistical significance.

- Discussion of the *implications* of these results, for example, *The results indicate that environmental factors influence levels of damage caused by the pest to fictitious pines in Sucrosia. Notably, levels of damage appear to be heavy at sites where the invasive weed fictional knotweed is abundant and the stand density high. This is consistent with previous reports* (followed by illustrative references regarding similar correlations affecting pine–pest interactions). The consistency (or otherwise) of any other correlations detected between damage and any other measured environmental variables with previous results then need to be mentioned (together with further illustrative references). In addition, possible explanations for the observed correlations should be described (with support, if available, from relevant references).
- Measurements of the frequencies of examined alleles, and any other calculated genetic parameters, such as within- and among-population diversity parameters, with confidence limits, followed by details of the correlations between damage levels, and each of the measured genetic variables (and their interactions), together with indications of their statistical significance.
- Discussion of the *implications* of the genetic *findings*, perhaps as follows: *These results show that genetic factors affect levels of damage. Notably, alleles* res2 *and* res4, *which are strongly associated with traits that influence resistance to insect pests, including bark thickness, the magnitude and distribution of resin ducts and the abundance of resin produced in response to insect herbivores* (followed by data showing the strength of the correlations, within and among populations, and the variance explained). *This is also consistent with previous reports* (followed by illustrative references regarding similar correlations affecting pine–pest interactions). Again, the consistency (or otherwise) of any other correlations detected between damage and any other measured genetic variables with previous results then need to be mentioned (together with further illustrative references). Similarly, possible explanations for the observed correlations should be described (with support, if available, from relevant references).
- As for a separate *Discussion* section, a combined *Results & Discussion* section should end with concluding remarks, which may or may not be presented in a separate section entitled *Conclusion* or *Conclusions*, and may be identical to the concluding remarks that would be included in a separate *Discussion* section, for example, for case study 2:

> The results indicate that both environmental and genetic factors influence levels of damage caused by *Hylobius fabricated* to fictitious pines in Sucrosia. Levels of damage appear to be heavy at sites where fictional knotweed is abundant and the stands are dense. In addition, the resistance alleles *res2* and *res4* appear to confer substantial resistance to the pest, in accordance with observations by several authors that homologous genes confer resistance to various insect pests in related pine species. Thus, if these threatened pines are to be conserved in Sucrosia (and potentially in other parts of the world), it is important to ensure that programs to manage remaining populations include clearing knotweed, thinning and planting saplings that carry *res2* and *res4* in re-planting programs.

As previously stressed, the suggestions for all sections of the papers presented thus far should be regarded as convenient frameworks or initial drafts, to which more detail can be easily added, but before this is done there are a number of other aspects to consider to help compose detailed drafts of the *Discussion* (or *Results & Discussion*), and other sections, to maximize chances of publication.

3.6.2 Further Reminders of Novelty

Regardless of the study, and the precise composition of the *Discussion*, it is important to consider the novelty and limitations (*focus*) of the study, as it is in other sections of the paper.

It is essential to highlight the novelty to remind readers that *The results show...* or *This is the first report of...* or *The results extend our knowledge of...* or *The data provide further indications of the importance of... this process/phenomenon.* For example, that urine of the imaginary wolf (and presumably volatiles within it) appears to be a potent deterrent of grazing by deer, and that both environmental and genetic factors are major contributors to resistance to the pest *Hylobius fabricated* in fictitious pines in Sucrosia.

3.7 Anomalies

Anomalies comprise a class of novelties (and/or limitations, depending on the context) that need to be especially carefully addressed. Milton Wainwright, of Sheffield University, often used to remark that 'anomalies should be cherished'. By this he meant that when results that conflict with previous results, or orthodox views, are obtained they should be carefully recorded and considered. In such cases people often try to find technical reasons (e.g., faulty instruments or inconsistencies in sampling or handling samples) that may explain the anomalies. For example, scientists using a Dobson Spectrophotometer detected the 'hole' in the ozone layer believed to be caused by Chlorofluorocarbons (CFCs) before it was detected by satellite data from the Total Ozone Mapping Spectrometer (TOMS) and the Solar Backscatter Ultraviolet (SBUV) instrument using more robust technology (NASA 2009), but they decided that their readings were due to instrumental error, thus missing a chance for an extremely high impact publication.

Like limitations, when one records anomalies, they need to be treated carefully, for several reasons. If they conflict with strongly entrenched views, they are clearly potentially highly important, but many referees who hold the entrenched views may raise fierce objections to them and seek to repress their publication. There may even be strong objections from non-scientific quarters. For this reason Galileo faced fierce opposition from many philosophers and clerics for supporting Copernicanism, and when he published his *Dialogue concerning two chief world systems* (1632) he was found to be *vehemently suspect of heresy* and spent the rest of his life under house arrest. Similarly, Darwin delayed publication of *On the Origin of Species* (1859) for several decades, because he knew it strongly conflicted with Creationist views of the universe and thus would cause an outcry in religious communities that hold such views.

In other cases, anomalies may conflict with results that your own research group has already published, suggesting that the previous results were incorrect or at least that more factors than those considered affect the observed phenomena. In such cases, ideally one would conduct further studies to resolve the apparent conflicts, but this may not be possible (e.g., if the project has finished and there are no funds for further experiments).

There are two possible strategies that can be followed in such situations. One is to present all of the *findings*, including the anomalies, and postulate reasons for the observed discrepancies with previous results. This strategy has the great virtue that it might provide good grounds for further grant applications. However, there is often a considerable risk that referees might decide that the postulated hypotheses are interesting, that they should be tested and the results of the tests should be incorporated in the paper before it is accepted (which, as mentioned, may not be possible). In such cases, the chances of publication may be increased by making the conflict the central focus of the paper (i.e., one could state in the *Abstract* and *Introduction* that *Previous studies have found that levels of cruciferoids* (a fictitious class of plant growth regulators) *form lateral gradients, across cambial meristems, and vertical, acripetal gradients in stems of crucifers. However, in the study presented here, although we found corroborative lateral gradients, basipetal rather than acripetal gradients were found in the stems. Possible reasons for the discrepancy with previous findings are discussed.*

The other strategy is to publish selective results (e.g., the corroborative results regarding the lateral gradient, perhaps accompanied by morphological observations or analyses of the effects of manipulating it using external applications of cruciferoids), ignoring conflicting results.

We know of cases in which both strategies have been successfully applied. The latter can be easily criticized for being unscientific, but alternatively it could be validly argued that no incorrect data are being presented, and it is highly possible that explanations for the anomalies will be forthcoming. Furthermore, if one acquires funding, one can re-examine the gradients and test possible explanations for the discrepancies (and hence publish another paper).

It should also be noted that interesting anomalies or novelties are sometimes 'buried' in papers, that is, mentioned very briefly, without emphasis. If a study has numerous novel features, this will not matter. However, for studies that lack much novelty, highlighting any novel (or strange, anomalous) features may be sufficient to attract a referee's or editor's attention sufficiently to decide that they should be published. For instance, if a study of factors associated with diabetes only detected the expected correlations between rates of the disease and weight, age, gender and so on, it would be unlikely to be published in a highly ranked journal. However, if the study also found that there was a consistent, weak but highly significant correlation between rates of the disease and the house numbers of the residences of the subjects, this should be highlighted, although the associated variation in rates was weak, since it would be so surprising. Hence, the paper would be more likely to be accepted, provided of course that the statistical analysis was sound.

3.8 A Strategy for Dealing with Major Limitations

As previously mentioned, in addition to novel aspects, every study has limitations, since it is impossible to collect fully representative samples, to address every factor and to include every possible control. The limitations cannot be ignored, or referees will insist that they are emphasized, but if they are highlighted too strongly the paper may be rejected. In one sense, the degree to which you can afford to describe the limitations varies inversely with the degree of limitations. In other words, if your study was extremely comprehensive, covering a very large range of potential factors with intensive and extensive sampling (and hence there were very small standard errors), the limitations could be discussed in great detail without worrying that this may lead to rejection of the paper. In contrast, if the limitations are severe (e.g., if you took very few samples or monitored just a few of a large class of pollutants at a few sites), they have to be mentioned *carefully*, and in such cases a strategy for addressing the limitations is required.

Usually, the best way to deal with severe limitations is to justify them in the *Introduction*, for example, by clearly stating why a few populations were *selected* for study or you *focused* on a few compounds (rather than *limited* the study to them). Generally, there is then no need to provide further justification of the limitations, provided that you do not make unwarranted extrapolations in the *Discussion*.

For example, if you only stated that your results indicate that the measured environmental and genetic factors identified in case study 2 affect interactions between *Hylobius fabricated* and *Pinus fictitious* in Sucrosia, referees are unlikely to object. However, if you stated that your results show that these factors are the main determinants of resistance of pines to insect pests, they would certainly object, because far more combinations of pines and pests would need to be examined to test the generality of the *findings*.

It is usually only desirable to mention the possibility that *findings* might be generally applicable if either you have examined a sufficient range of materials to support a novel assertion or you are planning to test the generality of the *findings*. If you have such plans, you could state, perhaps, the following:

> This study focused on populations of mosquitoes in southern Sucrosia. Thus, clearly populations of more species across a greater geographical range need to be examined to assess the generality of the findings. Nevertheless, the results indicate that the proposed control measures have considerable potential.

This strategy may be very useful for further grant applications, since if the paper is accepted, you can state that the potential validity of your hypothesis (that the *findings* might be generally applicable) has been accepted, because if referees do not object to the statement they are implicitly acknowledging the possibility. Hence, it may facilitate the publication of two or more papers rather than just one.

The importance of treating novelty and limitations carefully, in all sections of a paper, can be illustrated by a paper describing a process detected in Boreal forest ecosystems, which was submitted to *Nature*. A major problem was that the authors initially mentioned, repeatedly, that although the process they investigated had not been previously detected in natural environments, it had been shown to occur in laboratory settings, virtually asking the referees to reject the paper since the study was merely confirmatory. The paper was subsequently revised, before submission, by stressing when this was first mentioned that conditions were unrealistic in the previous laboratory experiments (concentrations of substrates were much higher than would ever be found in the field) and simply mentioning at other points that this was the first report of its occurrence in any ecosystem. These simple changes highlighted the novelty of the study, without completely ignoring its limitations, and the paper was accepted.

3.9 Figures and Tables

Figures and tables are generally used in the *Results* section, but they can also be useful for summarizing information elsewhere in a paper. For example, a figure in the form of a flow chart can sometimes be conveniently used to outline an analytical process in the *Materials & Methods* section or a table can be used to summarize the results of studies cited in the *Introduction*. Figures and tables can be particularly valuable where there is a strict word limit for the paper; however, many journals set a limit on the number of figures or tables that can be included, so they need to be used judiciously.

The material that needs to be incorporated in figures and tables should be carefully considered, because if too much information is included, it will be difficult to absorb it from them, but if there is too little, the referees and/or the journal's editor are likely to be annoyed (and hence more likely to reject the paper). Essentially, if the information you wish to present can be summarized in a few sentences, it should generally be included in the text in the appropriate section. However, if a long, complex paragraph, or more than one paragraph, would be required to describe the information in words, it should generally be presented in a table or figure. For instance, it would be absurd to describe differences in levels of large numbers of compounds found at a large number of sampling sites in words, rather than displaying the differences in a table or figure, or possibly a table in conjunction with Canonical Correspondence Analysis (CCA) score plots showing the patterns of differences between profiles detected at different kinds of sampling sites and arrows indicating correlations between the profiles and explanatory variables.

It can be tempting to include a large quantity of raw data in tables, but this should be avoided as it can lead to confusion and is usually unnecessary. If you feel that the availability of such data is an essential component of the paper, it is best to include it either as an appendix or as supplementary information, which could be made available via a website. As with the rest of a paper, consideration of the *rationale*, *objectives*, *what was done*, the *findings* and the *implications* of the study can help to identify appropriate information to include. Often, for example, *findings* can be summarized concisely and neatly in a table, and figures showing relationships between several variables can often visualize key *implications* of the results (e.g., that a proposed treatment is cost effective).

This is important because many people find patterns or relationships easiest to appreciate when data are presented visually, so figures can be particularly helpful when your results demonstrate patterns that are not easy to describe in words. Hence, figures can be particularly powerful for clearly illustrating your *findings*. For example, if there is a complex, non-linear relationship between two variables, the data can be presented as a scatter plot with a fitted regression line. Alternatively, when samples are clustered on the basis of a large number of variables, the data can be presented as an ordination plot (derived from appropriate multivariate analysis) with highlighted groupings.

Both tables and figures should be able to stand alone, that is, it should be possible to understand them in the absence of the associated text of the paper. Thus, legends should be succinct and informative, composed following the same procedure as for *Titles*, by concisely writing the *objectives* for the table or figure; for example:

> Map showing the locations of the selected populations/sampling sites

or

> PCA score plot showing the grouping of polyclic aromatic hydrocarbon profiles in locations with and without clusters of respiratory problems.

Many different table and figure styles and formats can be used, and preferences of journals vary widely in this respect. Therefore, it would be pointless (and potentially misleading) to give examples of 'good practice' in this guide. Instead, as for other aspects of your paper, you should consult the *Instructions for authors* provided by your target journal. Different journals also vary widely in the detail of instructions for tables and figures. The journal *Ecology*, for example, provides minimal guidelines (Ecological Society of America 2010), stipulating only the format of files that should be submitted. In contrast *Science* provides somewhat more detailed instructions regarding figures for both initial submissions (Science Magazine 2010a, b) and submissions following peer review (Science Magazine 2010c), but relatively few instructions about tables (Science Magazine 2010b). *Nature* stresses the need for simplicity in tables and figures as well as setting a limit on the number of words that can be included in a legend (Nature 2010). Both *Science* and *Nature* emphasize their page size, and this is an important consideration when planning tables and figures, since much detail can be lost when large items are reduced in size. In addition, many journals will not accept tables in landscape format.

It is also worth remembering that most journals will charge a fee (which can be substantial) for including color figures. Thus, if possible, ensure that your figures are meaningful in black and white. In cases where color is essential, funds will probably have to be found. However, some journals offer the option of presenting figures in black and white in the paper copy but in full color in the electronic version. Check the journal's position on this – it could be an important factor when deciding where to submit your paper.

In all cases, it is essential to follow journals' instructions, remembering that simplicity is likely to increase the impact of both tables and figures. To help gain a picture of what your target journal expects, consult some recent editions and base your design on figures and tables in them. Finally, as for the main body of the text, ensure that you submit your work in an acceptable electronic format and that your figures are at the correct resolution – you will not please the editors if they have to ask you to send your work again in a format they can read.

3.10 Reference Formatting Systems

As readers will be aware, there are diverse formats for references – the Harvard, Vancouver, APA (the American Psychological Association) and Turabian systems, to name but a few – and different journals have different preferences in this respect. Therefore, it is important to read, carefully, the *Instructions for authors*, and follow them (as it is for all sections of a paper). Although this is obvious, it is astonishing how often authors fail to do this, to the consequent annoyance of the journal's editor.

Detailed information is readily available on the internet about the different referencing styles (e.g., BMA 2006, University of Wales 2009). Indeed, Monash University (2010) provides a tutorial on the various referencing styles. Printed style guides also detail the formatting of specific referencing systems (JAMA 2007, Turabian 1996, University of Chicago Press 2003). A familiarity with the different styles is useful, but author preference is secondary – you should *always* follow the guidelines for your target journal. If in doubt, examine some papers published in the journal and follow the formatting in them.

When citing web pages, it is essential that you also note the date that you accessed the site, since such sources can be transitory. Since you may need to consult references again in the future, it is also best to save a copy of any web pages that you consult in case they are no longer available when you try to return to them. Here it is also worth noting that one should be discriminating when using the internet, relying only on trusted and verifiable sources.

Finally, it should be noted that it is essential to provide references to support all assertions, apart from those that are absolutely indisputable (e.g., the sun rises in the morning). Frequently, however, authors fail to do this. Notably, for instance, authors often state that increases in CO_2 concentrations are causing increases in global temperature. However, although there is a huge body of information supporting this hypothesis, and most scientists believe it, it is still disputed, and thus such assertions should be supported by one or more references. Similarly, one often sees assertions that various putative events occurred at certain points in geological history, without a supporting citation, although the timeline of geological events can only be inferred, and thus supporting evidence should always be cited.

References

BMA (2006) Reference styles: Harvard and Vancouver. http://www.bma.org.uk/library_medline/electronic_resources/factsheets/LIBReferenceStyles.jsp. Accessed 1 Sept 2010

Ecological Society of America (2010) Instructions for authors. http://esapubs.org/esapubs/AuthorInstructions.htm. Accessed 2 Sept 2010

JAMA (2007) AMA manual of style: a guide for authors and editors. Oxford University Press, New York, NY

Monash University (2010) Citing and referencing. http://www.lib.monash.edu.au/tutorials/citing/. Accessed 1 Sept 2010

NASA (2009) Ozone facts: history of the ozone hole. http://ozonewatch.gsfc.nasa.gov/facts/history.html. Accessed 1 Sept 2010

Nature (2010) Manuscript formatting guide. http://www.nature.com/nature/authors/gta/index.html#a5.7. Accessed 2 Sept 2010

Science Magazine (2010a) Preparing efficient figures for initial submission. http://www.sciencemag.org/about/authors/prep/prep_subfigs.dtl. Accessed 2 Sept 2010

Science Magazine (2010b) Preparing your initial manuscript. http://www.sciencemag.org/about/authors/prep/prep_init.dtl. Accessed 2 Sept 2010

Science Magazine (2010c) Preparing your art and figures. http://www.sciencemag.org/about/authors/prep/prep_revfigs.dtl. Accessed 2 Sept 2010

Turabian KL (1996) A manual for writers of term papers, theses and dissertations, 6th edn. University of Chicago Press, Chicago, IL

University of Chicago Press (2003) The Chicago manual of style: for authors, editors and copywriters, 15th edn. University of Chicago Press, Chicago, IL

University of Wales (2009) ReferenceStyles [sic]. http://www.wales.ac.uk/en/OnlineLibrary/StudySkills/ReferenceStyles.aspx. Accessed 1 Sept 2010

Chapter 4
Complex Studies

The previous hypothetical case studies were relatively straightforward, in that they could be described as single-part investigations, that is, the efficacy of wolf urine as a deer deterrent was examined in case study 1, while environmental and genetic factors influencing levels of attack by the pest *Hylobius fabricated* were simultaneously investigated in six fictitious pine populations in case study 2. Chapter 3 shows how simple statements regarding the *rationale*, *objectives* and so on of such relatively simple studies can be systematically used as frameworks to compose papers describing them. However, many studies are inherently more complex, since several kinds of materials or processes have to be investigated, either sequentially or in parallel. Writing a paper describing such an investigation is inherently more difficult. However, the same approach can be used; indeed, it is *more* important to construct a robust framework to ensure that no key information is excluded (or unnecessarily repeated) and that it can be readily understood. Thus, this chapter illustrates how the systematic approach can be applied when drafting papers reporting complex studies, beginning by showing how it could be applied to another hypothetical case study.

4.1 Hypothetical Case Study 3

This case study can be outlined as follows:

- At the start of the century, levels of respiratory problems, including pneumonia, chronic obstructive pulmonary disease (COPD) and asthma, were anomalously high in Averagetown, a medium-sized town with few other remarkable features. Such problems have several causes, including smoking, bacterial and viral infections, allergies, congenital diseases (e.g., cystic fibrosis) and various pollutants, inter alia particulate matter, sulfur dioxide, nitrogen dioxide, carbon monoxide, ozone and polycyclic aromatic hydrocarbons (PAHs).
- Possible sources of air-borne pollutants that could have been responsible included a large coal-fired power station and two waste incinerators, although all of them were equipped with appropriate scrubbers to remove pollutants from their emissions, and neither their monitoring equipment nor inspections by environmental health officers had provided any indications that these plants released high

J. Blackwell, J. Martin, *A Scientific Approach to Scientific Writing*,
DOI 10.1007/978-1-4419-9788-3_4,

levels of pollutants. There were also several factories in the area that could potentially have been responsible. Therefore, the local health authority commissioned a study to identify the cause(s) and recommend measures to address the problems.

- The commissioned researchers first examined hospitals' and doctors' records in Averagetown and two other medium-sized towns (Mediumtown and Standardtown) in the region to identify clearly the diseases and symptoms that were anomalously prevalent in Averagetown and assess the possibilities that the cluster of respiratory diseases in the town were attributable to a local outbreak of an infectious disease, a high frequency of any congenital disease known to cause such problems or pollutants (and if so identify the sources).
- A survey of clinical records indicated that neither infections nor congenital diseases were responsible for the high rates of respiratory problems in Averagetown. The researchers, therefore, sampled and measured levels of the pollutants mentioned before at several similar sites in Averagetown and the control towns, to determine whether they may have contributed to the respiratory problems, and to obtain more detailed information about the chemical profiles (relative proportions) of the pollutants potentially responsible. They also hoped to identify possible sources.
- Particulate matter and the gaseous pollutants were detected at levels greatly exceeding safety limits in Averagetown, but not the control towns.
- To identify the sources of the pollutants, the researchers (i) compared the profiles of the pollutants to those of known sources and (ii) collected samples at a range of sites in and around Averagetown to map the spatial distribution of the pollutants (and the temporal distribution of those that could be measured sufficiently quickly) and thus identify the area(s) and times that they were most abundant.
- The results of these parts of the study showed that the pollution profile did not closely match that of any known single, typical source. However, levels of some of the pollutants massively peaked several times a month, at night, and levels were highest close to one of the incineration plants.
- In further investigations, environmental health officers found that extremely toxic waste was being illegally (and highly profitably) incinerated and that the emissions were not being passed through the monitors (since they would have shown that the plant had been illegally operating) or the cleaning systems (which would have been massively overloaded).

Writing a *Title*, *Abstract* and other sections of papers describing such multi-part studies is more complex than writing corresponding sections of papers describing simpler investigations. Nevertheless, the process can be simplified using the systematic approach applied for case studies 1 and 2. Indeed, it is more important to adopt a systematic approach to ensure that the study is reported clearly, coherently and comprehensively (i.e., that key information is not omitted).

A complicating factor is that the *rationale* of such studies is intimately linked to the *objectives* and *findings*. For instance, if a local epidemic of a viral infection had been responsible for the respiratory problems examined in this case study, the subsequent monitoring and analyses would have been pointless. However, the *rationale*, *objectives* and *findings* for each part of the study can be conveniently expressed in a flow chart, which will be useful for tracking progress of the work and (eventually), writing progress reports and a paper (or papers) describing the study, as shown in Fig. 4.1.

A simple flow chart such as this essentially outlines the *rationale*, *objectives* and *main findings* of each part of the study in clear (chronological) order, since each step of the project after analysis of results of the survey of hospitals' and doctors' records is dependent on the results of preceding steps. So, the *rationale*, *objectives*, *what was done*, and the *results* of each step can be easily encapsulated using the simple statements in the flow chart (ignoring the possible outcomes that proved to be irrelevant), with appropriate modifications, as follows.

4.1.1 The Rationale, Objectives and Findings

- The incidence of respiratory problems was anomalously high in Averagetown.
- Therefore, records of hospitals and doctors in Averagetown were examined to quantify rates of the respiratory problems, and their symptoms were compared to those induced by known causes.
- Rates of respiratory problems were confirmed to be very high in Averagetown, and neither infections nor congenital diseases appeared to be responsible.
- Levels of pollutants were found to be far higher in Averagetown than in two control towns, and far higher than safe limits, but their profiles did not correlate with those of typical single sources.
- Higher spatial and temporal resolution monitoring indicated that an incinerator plant – which was intermittently, illegally burning highly toxic waste – was the major source of the pollutants.
- Further monitoring showed that levels of pollutants, and the incidence of the respiratory problems, declined when the illegal burning ceased.

Furthermore, since the *rationale*, *objectives* and *main findings* are interconnected, these brief statements can be simply re-worded to outline the *objectives*, *what was done* and the main *findings* of each step of the study, as required (see later).

Writing out the *rationale*, *objectives*, *what was done* and *main findings* in this manner may seem tedious. However, it can save huge amounts of time, because it provides a clear, coherent framework to follow when drafting each section of a paper describing the study. It may also seem difficult, but if you cannot outline a clear framework at this stage, it will be impossible to follow a clear order when adding all of the important details while writing a paper. Furthermore, having such a framework greatly helps to avoid the omission of key information when writing the paper, as illustrated in following parts of this guide, first, by showing how it can be used to draft each section of a paper describing case study 3.

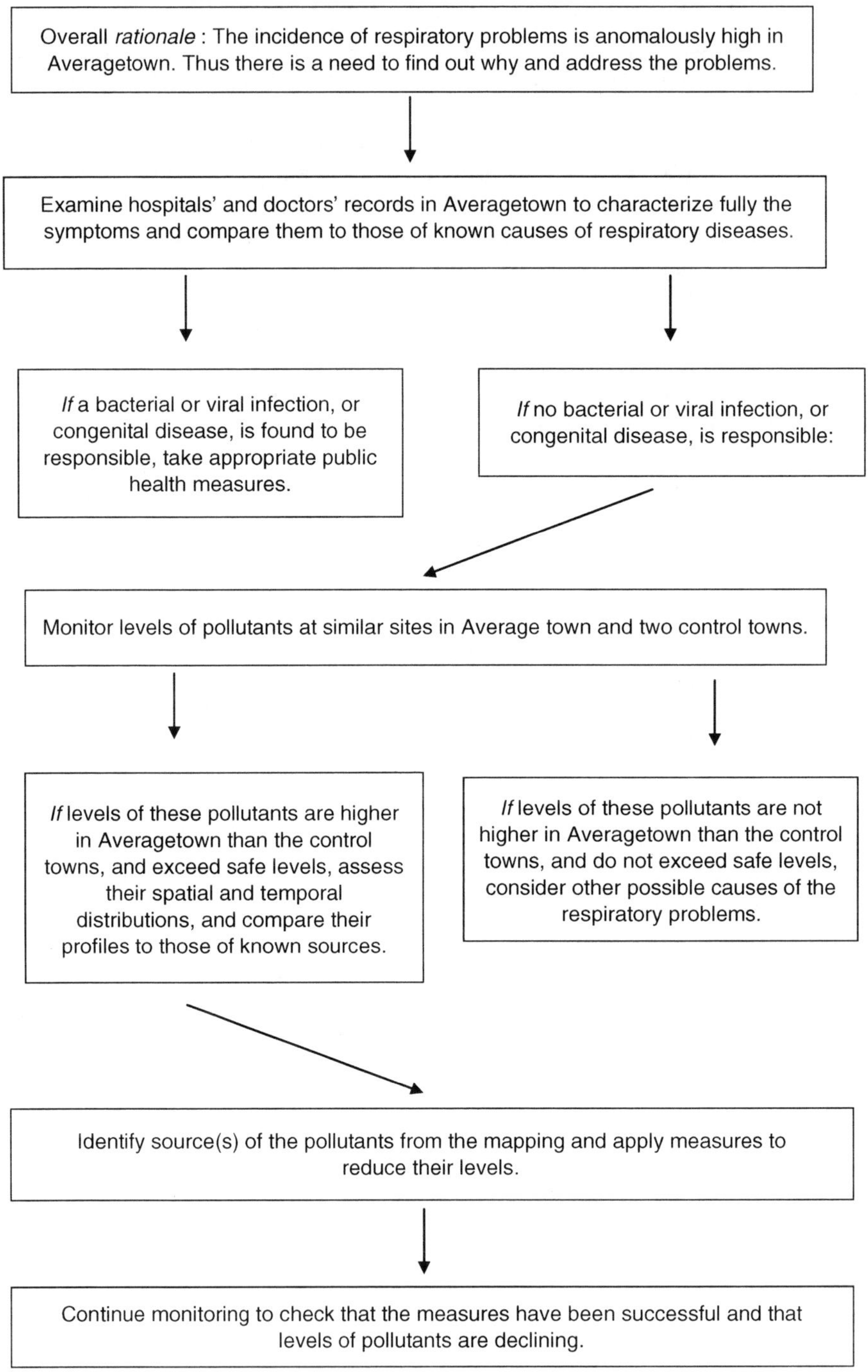

Fig. 4.1 Flow chart showing the *rationale*, *objectives* and *possible outcomes* of case study 3

4.1.2 Title and Abstract

As for the previous hypothetical studies, a suitable *Title* can be based on the main *findings* or *implications*:

> High levels of respiratory problems in Averagetown were due to illegal combustion of highly toxic waste in an incinerator.

Similarly, as for the previous studies, an *Abstract* can be composed by briefly describing the overall *rationale*, and then the *objectives*, *what was done*, the *findings* and the *implications*, *but not in quite the same way as for the simpler studies.* Since these aspects are intimately linked in a complex study, if the *rationale, objectives* and *findings* for each step were sequentially outlined, the resulting *Abstract* would be extremely long and repetitive. Therefore, sentences encapsulating the progression of the study, incorporating these elements, should be written instead. This can be done by re-wording, and linking, the summarizing statements, perhaps as follows:

> The incidence of respiratory problems was anomalously high in Averagetown at the start of the century. To elucidate the cause(s) of these problems, we first examined records of hospital and doctors in the town to quantify the incidence of the problems and characterize the symptoms fully. This analysis confirmed that rates of the problems were very high in Averagetown, and indicated that this was not due to smoking, infection (bacterial or viral) or any diagnosed congenital disease. We then monitored levels of pollutants known to cause such problems (including particulate matter, sulfur dioxide, nitrogen dioxide, carbon monoxide, ozone and PAHs) at Averagetown and, at similar sites, in two similar control towns in the region (Mediumtown and Standardtown). Levels of all these pollutants were found to be markedly higher in Averagetown than in the control towns, and exceeded guideline limits by several orders of magnitude. Spatial and temporal mapping of the pollutants in Averagetown indicated that their main source was a waste incinerating plant, which was subsequently found to be illegally incinerating highly toxic waste several times a month. This practice has now ended and further monitoring showed that both levels of pollutants and the incidence of the problems are declining.

All that would then remain to be done would be to add a few appropriate details regarding (for instance) the Odds Ratios of the respiratory problems, the methods used to measure the pollutants and their measured and guideline limit concentrations.

4.1.3 Introduction

Composing *Introductions* for papers describing the previous two case studies would also have been fairly straightforward, because similar populations, processes or phenomena were examined, and the same treatments were applied or variables were measured in each case. Writing *Introductions* for papers describing more complex studies, in which several populations, processes or phenomena were investigated, sequentially or in parallel, is more challenging, since relevant aspects of each population, process or phenomenon (and the pros and cons of the methodology that could be applied in each case) have to be covered.

However, approaching the task systematically can be highly beneficial. Indeed, it is even *more* important for complex studies than for simple studies, in order to maintain the narrative flow and avoid unnecessary repetition and omissions. This can be conveniently done by using the summarizing statements, with appropriate re-wording in the form of a *rationale*:

- The incidence of respiratory problems was anomalously high in Averagetown.
- Therefore, records of hospitals and doctors in Averagetown were examined to quantify rates of the respiratory problems, and their symptoms were compared to those induced by known causes.
- Rates of respiratory problems were confirmed to be very high in Averagetown, and neither infections nor congenital diseases appeared to be responsible.
- Levels of pollutants were found to be far higher in Averagetown than in two control towns, and far higher than safe limits, but their profiles did not correlate with those of typical single sources.
- Higher spatial and temporal resolution monitoring indicated that an incinerator plant was the major intermittent source of the pollutants. This plant was found to be illegally burning highly toxic waste.
- Further monitoring showed that levels of pollutants, and the rates of the respiratory problems, declined when the illegal burning ceased.

Using these statements as a framework, and adding details from fuller notes made during the study, a draft of the *Introduction* should start by expanding the initial statement:

> The incidence of respiratory problems (including, pneumonia, chronic obstructive pulmonary disease and asthma) was believed to be anomalously high in Averagetown at the start of this century (essential references). Thus, there was an urgent need to identify the causes and possible measures to address the problems.

It could then continue as follows:

> Records of hospitals and doctors can be very valuable for assessing rates of diseases, their symptoms and diagnoses (followed by references and discussion of other methods that could be used and their pros and cons). Known causes of such problems include bacterial and viral infections, smoking, and various pollutants (followed by brief descriptions of associated symptoms, with illustrative references).
>
> If examination of records indicates that infections, smoking and congenital diseases are not major causes of the problems, then pollutants are likely causes, and this possibility can be assessed by appropriate measurements of levels and profiles of potential pollutants in the affected area and one or more control areas. Possible measuring techniques include (followed by outlining various kinds of passive and active sampling that could be used, and their advantages and disadvantages, highlighting the merits of those used, with illustrative references). If levels of pollutants are found to be dangerously high in the affected area, higher resolution (spatial and temporal) sampling can be used to identify possible sources of the pollutants and potential remedial measures (followed by illustrative examples from the literature).

Having done all this, the *Introduction* should conclude by outlining the *objectives* of the study, and *what was done*, as for the previous studies. However, since these aspects of a complex study are inter-linked with the results, some of the main *findings* also need to be mentioned at this point, to show why the sequence of investigations was followed. These inter-linked aspects can be simply formulated in the form of concluding remarks by re-wording, and linking, appropriate statements summarizing the study:

> In the study presented here, records of hospitals and doctors in Averagetown were first examined to quantify rates and symptoms of respiratory problems in the town, and eliminate some potential causes. The rates were confirmed to be anomalously high, and neither infections nor congenital diseases appeared to be responsible. Therefore, levels of pollutants were measured in Averagetown and two control towns. They were found to be dangerously high in Averagetown, so the monitoring was extended, with higher spatial and temporal resolution, in Averagetown, and a major source was found to be a waste incinerator that was intermittently burning highly toxic waste illegally.

Again, an alternative to this kind of construction that could be adopted (*but should be avoided*) would be to incorporate, briefly, *what was done* in each part of the study, and the main *findings*, within the main part of the *Introduction*. For instance, in this case, after stating that the incidence of respiratory problems was believed to be anomalously high in Averagetown at the start of this century, and that hospitals' and doctors' records can be very valuable for determining rates of diseases and their symptoms, the *Introduction* could continue:

> In the presented study, examination of records indicated that rates of disease were anomalously high in Averagetown, but neither infections nor congenital diseases appeared to be responsible for the respiratory problems.

Similarly, after stating (i) that if records indicate that neither infections nor congenital diseases are the cause of such problems, pollutants are likely causes, (ii) that this possibility can be assessed by measuring levels and profiles of potential pollutants in the affected area, and (iii) outlining possible measuring techniques, the *Introduction* could state:

> Therefore, levels of pollutants were measured in Averagetown and two control towns, and they were found to be dangerously high in Averagetown.

As previously mentioned, this kind of construction inevitably leads to repeated jumps from general considerations to *what was done* and back to general considerations, thereby introducing unnecessary repetition and disrupting the narrative flow. In addition, if this approach is adopted, some of the information that will be repeated again in the *Materials & Methods* and *Results* sections is inevitably presented several times in the *Introduction*, adding further unnecessary repetition.

Thus, to maintain narrative flow and avoid unnecessary repetition, it is generally *much* better to outline, fully, all of the key aspects of the context of the study, starting with the most general aspects followed by more specific aspects, *and then* conclude the *Introduction* by outlining the specific *objectives* of the study and *what was done*.

4.1.4 Materials & Methods

For this study, a *Materials & Methods* section can be composed using the summarizing statements, re-worded in *what was done* format:

- Records of hospitals and doctors in Averagetown were examined to quantify rates of the respiratory problems, and their symptoms were compared to those induced by known causes.
- Levels of pollutants were then measured in Averagetown and two control towns, and their profiles were compared to those of typical sources.
- Higher spatial and temporal resolution monitoring was then undertaken to identify potential sources of pollutants.
- Monitoring was continued after the illegal burning ceased to check that levels of pollutants, and the rates of the respiratory problems, were declining.

Adding further details from notes taken during the study, a draft of the section should begin by describing the hospital and doctors' records examined, in terms of the population, dates and types of problems covered (if, for instance, patients with some potentially relevant congenital diseases were diagnosed and treated elsewhere, this should be mentioned, together with any measures applied to account for gaps in the data). It should then describe, in order:

- The methods used to extract information regarding the incidence of respiratory problems, and their symptoms, from the hospital and doctors' records, and the sources of any comparative regional, national or international data.
- The statistical methods used to compare rates of observed syndromes with those in other areas (with illustrative references for the methods).
- Methods used to compare observed syndromes that were anomalously prevalent in Averagetown to those of known causes of respiratory problems, in order to identify possible causes and exclude others.
- The numbers and locations of pollutant sampling sites (and the criteria used to select them), in both the initial and higher-resolution monitoring campaigns.
- The methods, equipment and settings (with essential references) used to sample and analyse the pollutants, in both campaigns.
- The methods and software used to compare profiles of pollutants to those of known sources, and pin-point the putative source(s), and the sources of comparative data (e.g., limit values specified by relevant authorities), again with essential references.
- Finally, the length of time that monitoring was continued to check if levels of the pollutants and rates of the respiratory problems were declining following cessation of illegal waste burning at the incinerator.

4.1.5 Results

For this complex study, the summarizing statements can be used, in *findings* format, to frame the *Results* section, in the same manner that they were used to frame the *Materials & Methods* section in *what was done* format (thus ensuring that results are presented in a corresponding sequence):

- Examination of the records confirmed that rates of the respiratory problems in Averagetown were very high, and neither infections nor congenital diseases appeared to be responsible.
- Levels of particulate matter, sulfur dioxide, nitrogen dioxide, carbon monoxide, ozone and PAHs were far higher in Averagetown than in the control towns, they far exceeded safe limits, and their profiles did not match those of known single sources.
- Higher spatial and temporal resolution sampling showed that a waste incinerator was the probable source of the pollutants.
- Levels of pollutants, and rates of the respiratory problems, declined after the incinerator plant stopped illegally burning toxic waste.

Hence, the *Results* section should sequentially describe:

- Calculated rates of the respiratory problems in Averagetown and the degrees to which they exceeded rates in other parts of the region and elsewhere (with confidence limits).
- Measured levels, ranges and profiles of pollutants at the sampling sites (together with indications of confidence levels, spatial variations and comparisons to official safe levels, with essential references) in each of the towns, in the low-resolution sampling campaign.
- Measured levels, ranges and profiles of pollutants at the sampling sites (together with indications of confidence levels, spatial and temporal variations and comparisons to official safe levels) in Averagetown, in the high-resolution sampling campaign.
- Measurements showing the levels of the pollutants and rates of respiratory problems had declined following cessation of the illegal burning.

4.1.6 Discussion

As for the simpler case studies, a draft *Discussion* for a paper describing this study should begin by briefly recapitulating the overall *rationale*, using the summarizing statements:

> As mentioned earlier, the incidence of respiratory problems (including pneumonia, COPD and asthma) was believed to be anomalously high in Averagetown at the start of this century (essential reference). Hence, there was an urgent need to identify the causes and possible measures to address the problems.

It could then continue by sequentially outlining the *rationale* (very briefly) for each of the main *findings* at each stage of the investigation, in the order they were presented in the *Results* section, and discussing how they compare with results of previous studies and their implications, as follows:

> Possible reasons for the high incidence of respiratory problems in Averagetown were examined by, first, examining hospital and doctors' records. The records confirmed that rates of respiratory problems were very high in Averagetown (followed by a brief summary of data showing how much higher they were in the town than in other places in the region and elsewhere, with illustrative examples from the literature). Causes of these diseases include smoking, various infections and congenital diseases (with illustrative examples from the literature). However, the analysis of the records indicated that rates of the congenital diseases considered, smoking and infections were no higher than national averages (followed by illustrative data and references). Thus, these potential risk factors were excluded as probable causes of the unexpectedly high rates of respiratory problems.
>
> Therefore, attention turned to the possibility that pollutants may have been responsible, since several classes of pollutants are known to cause respiratory problems (followed by examples from the literature). To assess this possibility, levels of particulate matter, sulfur dioxide, nitrogen dioxide, carbon monoxide, ozone and PAHs were monitored in Averagetown and two control towns (Mediumtown and Standardtown), and the profiles of the pollutants were evaluated. The results of this monitoring campaign showed that levels of all of these pollutants were several orders of magnitude higher than proscribed levels in Averagetown, but not the control towns (with illustrative data, significance values and essential references).
>
> The probable sources of pollutants can often be identified by comparing observed profiles to those of known sources (followed by illustrative examples). However, in this case, the profiles obtained from sampling in Averagetown did not closely match those of any known sources, for example, relative levels of PAH congeners did not match those associated with traffic or burning coal (with illustrative references).
>
> Therefore, in an attempt to pinpoint the source(s) of the pollutants, the spatial and temporal resolution of the sampling in Averagetown was increased. Their levels were found to peak massively several times a month, during the night, and the probable source appeared to be a waste incinerating plant. In further investigations, environmental health officers found that highly toxic waste was being incinerated illegally at this plant, with the emission monitoring systems by-passed, to avoid records of the activity being logged, and the scrubbing systems off-line, to avoid them being massively overloaded. This activity has now ceased, and both levels of the pollutants and rates of the respiratory problems in Averagetown have declined substantially.

4.1.7 Conclusion

This can be based on the final statements extracted from the flow chart (Fig. 4.1).

> In conclusion, the results show that pollutants were largely responsible for the high rates of respiratory problems in Averagetown, and that a major source of the pollutants was the illegal burning of highly toxic waste at a local incinerator. The study also demonstrates the power of high-resolution monitoring for pinpointing sources of pollutants.

It should be noted that some referees or editors, who prefer use of the 'personal voice', would dislike the first sentence of this conclusion and could insist that it be changed to *We conclude that pollutants...* (although, on 15 August 2010, Google only returned 647,000 hits for '*We conclude that*' + *gene* + *PCR*, compared to 3 million for '*The results show*' + *gene* + *PCR*, 2.6 million for '*The results suggest*' + *gene* + *PCR*, and 2.4 million for '*The results indicate*' + *gene* + *PCR*).

Use of personal, impersonal, passive and active voices is considered in detail later (Section 5.3).

4.2 Hypothetical Case Study 4

To show that the described procedure can be applied even to very complex studies, let us consider the last hypothetical case study, which can be outlined as follows:

- Recent large-scale molecular screening indicates that an xx mutation in receptors for the hormone Human Growth Factor Fictitious (HGFF) expressed in certain tissues, which reduces their sensitivity to the hormone, may be responsible for Anon's disease, a human growth deficiency syndrome.
- In addition, three of a class of small molecular weight drugs developed by a pharmaceutical company, designated X189-X191, have slight therapeutic effects on Anon's disease.
- If the mutation of the receptors is a cause of Anon's disease, X189-X191 may act by enhancing the affinity of the receptors for the hormone and/or binding of the enzyme, HGFF signaling kinase (HGFFs kinase), which triggers subsequent signaling events after the hormone has bound to the receptors. If so, variants of X189-X191 could potentially be rationally designed that are capable of restoring the activity of the receptors more completely and thus provide much more effective treatment of the disease.
- To test these hypotheses, assessments of the effects of the mutation on the affinity of the receptors for the hormone, and the affinity of the hormone-receptor complex for the kinase, are required.
- If the mutation proves to reduce either of these affinities, it is then necessary to assess whether or not compounds X189-X191 partially restore affinity.
- In addition, structural analyses of the wild type (normal) and mutated forms of the receptor, alone and in complex with the hormone, both in the presence and absence of compounds X189-X191, are required to determine the effects of the mutation on the three-dimensional structure of the receptor.
- In parallel to all this, animals should be identified or developed that have Anon's disease ('animal models') to check whether they carry analogous mutations, thus corroborating the hypothesis that the mutation causes the disease and providing subjects that could be used in pre-clinical trials of new variants of compounds X189-X191.
- Finally, if the hypothesis is confirmed, and variants of compounds X189-X191 (or other drug candidates) can be designed that enhance the activity of the mutated receptor, and animal models can be identified or developed that carry analogous mutations, then the toxicity and potential therapeutic efficacy of the variants should be tested.

Clearly, it is impossible to describe the *rationale* for such a study in a single brief statement. However, as for case study 3, it can be described in a series of short

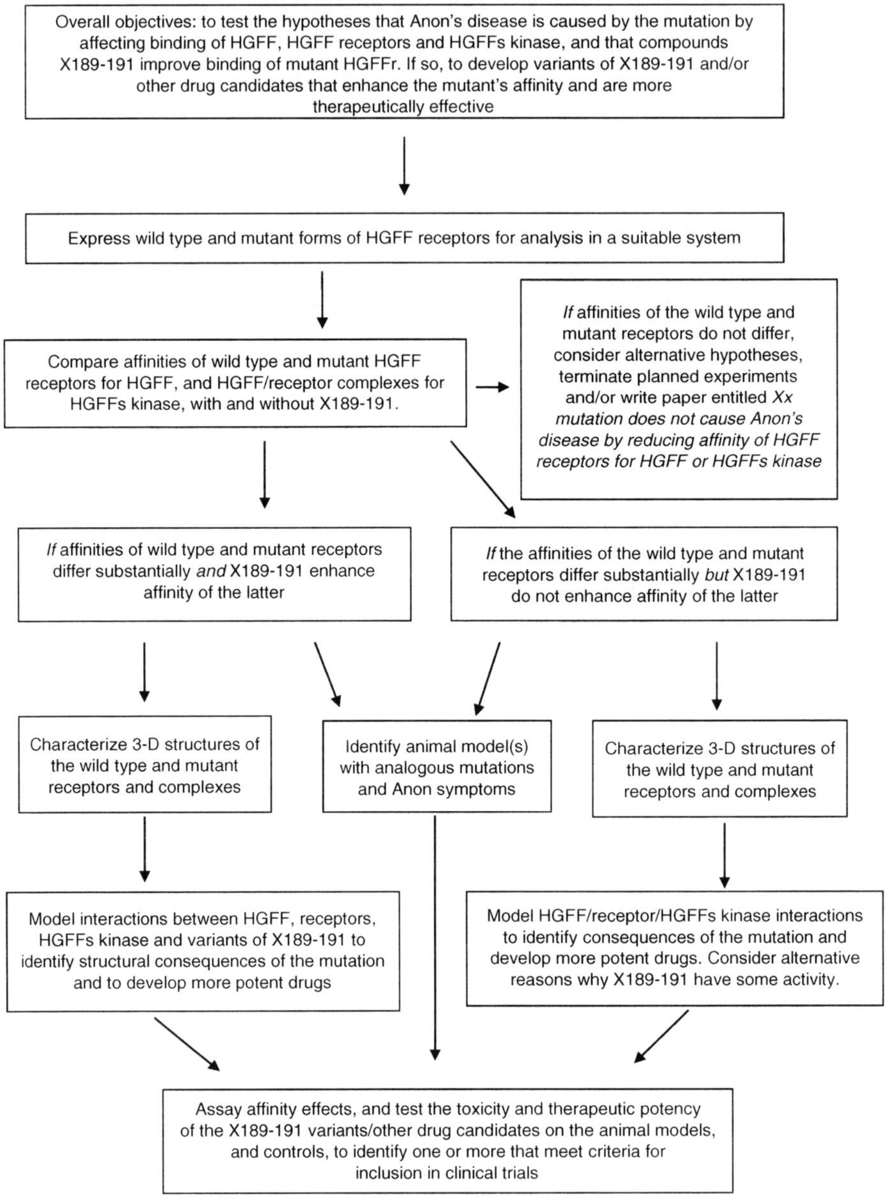

Fig. 4.2 Flow chart showing the *rationale*, *objectives* and *possible outcomes* of case study 4

statements that follow a coherent sequence (e.g., the mutation must first be shown to influence the binding affinity of the receptors significantly before doing most of the subsequent work, otherwise it would be pointless). Thus, as in case study 3, the *rationale*, *objectives*, *what was done* and the *findings* are intimately connected and can be conveniently expressed in a flow chart, which will be useful for constructing a more detailed summary, tracking progress of the work, writing progress reports and, eventually, drafting a paper (or papers) describing the study. A flow chart for case study 4 is presented in Fig. 4.2.

Like the flow chart prepared for case study 3, this one essentially outlines the *rationale* of specific parts of the study, the *objectives*, *what was done* and the *findings*, for various permutations of outcomes, since each step of the project after comparison of the affinities of the wild type and mutant receptors for HGFF and HGFFs kinase is dependent on the results of the preceding steps. Hence, as for case study 3, it provides a brief, coherent framework for writing each section of a paper describing the study, regardless of the outcome.

In practice, several papers would probably be written during the course of such a massive study (in which case the previously described procedure for writing a *Title* and *Abstract*, etc., could be used for each paper). However, it is possible that it may be presented in a single paper, for instance, if it was being undertaken by the company that developed compounds X189-X191 and they wanted to keep the results confidential until more potent variants had been developed and were ready or nearly ready to enter clinical trials. In addition, a final report for the project would have to detail all of its stages.

Hence, we will illustrate how a paper or report describing such a complex study could be written, starting from the simple statements regarding the *rationale*, *objectives* and *findings* in the flow chart, assuming that the mutation affects the affinity of the receptors for both HGFF and HGFFs kinase, that X189-X191 do partially restore the activity of the receptor, that a suitable animal model (designated Anon rat) was identified that carries the mutation and presents Anon's disease symptoms, that the structural determinations were successful and that five more potent variants of X189-X191 (designated XE1-XE5) were identified, two of which met toxicity criteria for inclusion in clinical trials.

As for case study 3, a step that can save huge amounts of time is to re-phrase relevant parts of the flow chart to generate brief statements encapsulating the *rationale*, *objective*, *what was done* and *findings* for each step of the study, which can be subsequently re-worded to obtain corresponding statements regarding the *objectives*, *what was done* and the *findings* separately, yielding a coherent framework for each section of the paper, as follows.

4.2.1 The Rationale, Objectives and Findings

- It has been postulated that Anon's disease is caused by a mutation of the HGFF receptor (HGFFr) that affects binding of HGFFr to HGFF and/or HGFFs kinase and that compounds X189-X191 improve binding of mutant HGFFr.

- To test these hypotheses, and develop more potent variants of X189-X191 and/or other drug candidates, wild type and mutant forms of the receptor were obtained for analysis by expressing them in a suitable system.
- The affinities of the wild type and mutant forms of the receptor for HGFF, and hormone-receptor complexes for HGFFs kinase, in the presence and absence of compounds X189-X191, were then compared.
- The mutation was found to severely reduce the affinities of the receptor, therefore 3-D structures of the wild type and mutant receptors, and complexes, were determined to identify structural and functional consequences of the mutation.
- Furthermore, X189-X191 significantly enhanced the affinity of HGFFr for HGFF and HGFFs kinase, so interactions between HGFF, the mutant receptor, HGFFs kinase and variants of X189-X191 were then modeled to develop more potent drugs.
- In addition, an animal model with analogous mutations that presented Anon symptoms was identified.
- Finally, the affinity effects were assayed, and the toxicity and therapeutic potency of several X189-X191 variants were tested using the animal model and controls, to identify ones that are both more potent than the lead compounds and meet criteria for inclusion in clinical trials.

Again, specific aspects of the study can be summarized by simply re-wording these statements, and using them to draft each section of a paper, as follows.

4.2.2 Title and Abstract

Let us first consider how these brief summaries can help to write a *Title* and *Abstract*. Clearly, no concise title could incorporate all the *findings*, but as for the previous hypothetical studies, a suitable title can be based on the main *findings* or *implications*, as follows:

> Xx mutation in HGFF receptors affects HGFF binding thus causing Anon's disease, which can be treated by variants of X189-X191 in a rat model.

In addition, as for the other complex case study (3), an *Abstract* can be composed by re-wording and linking the statements summarizing the study, adding appropriate details, as follows:

> Recent large-scale molecular screening indicates that an xx mutation in Human Growth Factor Fictitious (HGFF) receptors expressed in certain tissues, which reduces their affinity for the hormone and/or HGFFs kinase, may be responsible for the human growth deficiency syndrome Anon's disease. In addition, the small molecular weight drugs X189-X191 have slight therapeutic effects toward Anon's disease. To determine whether xx mutation causes Anon's disease by reducing the affinity of HGFF receptors for HGFF and/or HGFFs kinase, and if X189-X191 act by enhancing the activities of the receptors, we expressed wild type and mutant receptors in ww and compared their affinities for HGFF and HGFFs kinase, in the presence and absence of X189-X191 in yy assays. We then characterized the 3-D structures of the wild type and mutant receptors by mm spectroscopy and nn modeling, and designed variants of X189-X191 (designated XE1-XE5) that restore the activity of

the mutant more effectively, by stabilizing a non-native conformation that presents active ligand-binding sites, using zz software. In addition we examined potential animal models and identified a rat line, designated Anon, which carries the mutation in addition to displaying Anon symptoms. Qq toxicity tests showed that XE1 and 4, but not XE2, 3 or 5, met standard toxicity criteria for clinical trials and substantially alleviated Anon symptoms in Anon rats. Hence, the results confirm the hypotheses and indicate that XE1 and 4 have considerable therapeutic potential for treating Anon's disease.

4.2.3 Introduction

The *Introduction* of a paper describing this complex study can also be framed by re-wording the brief summary of the study in the form of a *rationale*:

- It has been previously postulated that Anon's disease is caused by a mutation of HGFFr that affects binding of HGFFr to HGFF and/or HGFFs kinase, and that compounds X189-X191 improve binding of mutant HGFFr.
- To test these hypotheses, and develop more potent variants of X189-X191 and/or other drug candidates, wild type and mutant forms of the receptor were obtained for analysis by expressing them in a suitable system.
- The affinities of the wild type and mutant forms of the receptor for HGFF, and hormone-receptor complexes for HGFFs kinase, in the presence and absence of compounds X189-X191, were then compared.
- The mutation was found to severely reduce the affinities of the receptor, therefore 3-D structures of the wild type and mutant receptors, and complexes, were determined to identify structural and functional consequences of the mutation.
- Furthermore, X189-X191 significantly enhanced the affinity for HGFF and HGFFs kinase, so interactions between HGFF, the mutant receptor, HGFFs kinase and variants of X189-X191 were then modeled to develop more potent drugs.
- In addition, an animal model with analogous mutations that presented Anon symptoms was identified.
- Finally, the affinity effects were assayed, and the toxicity and therapeutic potency of several X189-X191 variants were tested using the animal model and controls, to identify ones that are both more potent than the lead compounds and meet criteria for inclusion in clinical trials.

Following these statements, the *Introduction* should begin by describing Anon's disease and the problems it causes, to establish the importance of the study.

It should then continue by:

- Outlining previous hypotheses (if any) regarding the pathogenesis of the disease, the reasons that they failed to describe all of the symptoms and/or were unlikely to be valid, and the recent hypothesis that the cause of the disease may be the xx mutation in HGFF receptors (all with essential references).

- Citing evidence that X189-X191 slightly ameliorate the disease, and the hypothesis that these compounds may act by enhancing the activities of the mutant receptors.
- Stating that in order to characterize interactions between receptors and the molecules they interact with (ligands), it is essential to obtain them in sufficient amounts, in sufficiently pure and native states, then describing systems that can be used to do this, and their advantages and disadvantages (with illustrative references).
- Describing assays that could be used to assess the affinity of the receptors for their ligands and to assess their affinity (with illustrative references).
- Mentioning methods and software (with essential references) that can be used to characterize the structure of receptors and complexes they form, and to design variants of drug candidates that have greater potency than lead compounds (noting their advantages, especially for the methods used, and disadvantages).
- Highlighting the value of using animal models to explore the pathogenesis of diseases, and the methods used to identify models that carry specific mutations and display symptoms of investigated diseases, with illustrative examples from the literature.
- The context-setting of the study could then finish by describing the need to assess the toxicity and therapeutic efficacy of candidate drugs (with illustrative references).

Note, as for the previous case studies, it is not necessary to state that any of the techniques mentioned in this part of the *Introduction* were used, it is only essential to mention those that were used and highlight their advantages to justify their application.

Once the *rationale* for the study has been described, as outlined earlier, the *Introduction* can be concluded by outlining the *objectives* and *what was done*, by merging and re-wording the summarizing statements, adding sufficient detail to show why each part of the study was performed:

> In the study presented here, to obtain wild type and mutant forms of the receptor for analysis, we expressed them in ww. We then compared their affinities for HGFF, and hormone-receptor complexes for HGFFs kinase, in the presence and absence of compounds X189-X191, in yy assays. The results showed that the mutation adversely affected the affinity of the receptors for both HGFF and HGFFs kinase, and the presence of X189-X191 partially restored the activity of mutant receptors by stabilizing a non-native conformation that presents active ligand-binding sites. Therefore, we characterized 3-D structures of the wild type and mutant receptors and complexes, using mm spectroscopy and designed (using zz software) five variants of X189-X191, designated XE1-XE5, that appeared to be more potent than the lead compounds in simulations. The interactions between HGFF, mutant receptors, HGFFs kinase and the X189-191 variants were then empirically characterized using hh hybrid assays, to confirm that XE1-5 improved the affinities of the receptors. In parallel with these efforts, we characterized several potential model animals and identified one (Anon rat) that both carries the mutation and presents Anon symptoms. Finally, we tested the toxicity of XE1-XE5, in yy assays and their therapeutic effectiveness toward the animal model.

4.2.4 Materials & Methods

As for case study 3, the *Materials & Methods* section for this study can be conveniently drafted using the statements summarizing the study, in *what was done* format, as a framework:

- Wild type and mutant forms of the receptor were expressed.
- Affinities of the wild type and mutant forms of the HGFF receptor for HGFF, and hormone-receptor complexes for HGFFs kinase, in the presence and absence of compounds X189-X191, were determined.
- 3-D structures of the wild type and mutant receptors and complexes were characterized.
- Interactions between HGFF, HGFFr, HGFFs kinase and variants of X189-X191 were modeled and empirically characterized in order to develop more potent drugs.
- An animal model with an analogous mutation, presenting Anon symptoms, was identified.
- Finally, the toxicity and therapeutic potency of X189-X191 variants found to increase the affinity of the mutant receptor were assessed, using the animal model and controls.

Expanding these statements, a draft of the *Materials & Methods* section of the paper should begin by describing how the receptors were expressed (including details of any sequencing, and/or sources used to acquire sequences, vectors, the expression system and transformation protocols, with essential references).

It should then describe, in order:

- Assays used to characterize the affinities of the wild type and mutant forms of the receptor, and complexes, in the presence and absence of X189-X191, with essential references.
- The methods, and software, used to model the interactions between the molecules, and to design more potent variants of X189-X191 and/or other drug candidates, with essential references.
- The techniques used to identify Anon rat as an animal model carrying the mutation and displaying Anon symptoms, with essential references.
- Finally, the assays used to assess the affinity effects, and the toxicity and therapeutic potency of X189-X191 variants, with essential references.

4.2.5 Results

For this complex study too, a summary of the results obtained by re-wording the statements in the flow chart (Fig. 4.2) describing the study can be used to frame a *Results* section, in the same manner they were used to write the *Materials & Methods* section, but in *findings* format:

- Wild type and mutant forms of the receptor were expressed.
- Affinities of the wild type and mutant forms of the HGFF receptor for HGFF, and hormone-receptor complexes for HGFFs kinase, in the presence and absence of compounds X189-X191, were determined.
- 3-D structures of the wild type and mutant receptors and complexes were characterized.
- Interactions between HGFF, HGFF receptors, HGFFs kinase and variants of X189-X191 were modeled in order to develop more potent drugs, and the drugs were found to stabilize a non-native HGFF receptor conformation that presents active ligand-binding sites.
- An animal model with an analogous mutation was identified.
- Finally, the toxicity and therapeutic potency of X189-X191 variants found to increase the affinity of the mutant receptor were assessed, using the animal model and controls.

Therefore, adding details from notes on the *findings*, the *Results* section should describe, in order, results of the following:

- analyses showing that the receptors had been expressed correctly by the selected expression system
- the affinity assays and the analyses of the effects of X189-X191 on the receptor's affinities
- the structural analyses
- the screenings used to identify an appropriate animal model, and analyses used to confirm that Anon rat was suitable
- modeling of the interactions between HGFF receptors and the other molecules
- the toxicity and therapeutic potency assays of the X189-X191 variants.

4.2.6 Discussion

As for all of the previous studies, a drafted *Discussion* for a paper reporting this study should begin by recapitulating the first part of the *rationale*, using the summarizing statements, and notes taken during the study as a guide:

> As mentioned in the Introduction, various hypotheses have been postulated in attempts to explain the pathogenesis of Anon's disease, most of which have serious inconsistencies with observed symptoms (followed by illustrative examples from the literature). However, the recent hypothesis that the disease may be caused by xx mutation in HGFF receptors seems to be more concordant with the symptoms (essential reference). The results of the affinity tests presented here, using wild type and mutant receptors expressed in ww, strongly corroborate the hypothesis, since the dissociation constants for HGFF/receptors and HGFF/receptor/HGFFs kinase complexes were several orders of magnitude higher for complexes with the mutant receptor than for complexes with the wild type receptor. Furthermore, they show that compounds X189-X191 partially restore the affinity of the mutant receptor, reducing these dissociation constants several-fold (followed by a summary of the relevant data).

Then, as before, it should sequentially outline each of the main *findings* (in the order the corresponding methodology was presented in the *Results* section), and discuss how they compare with the results of previous studies, and their implications:

> The following structural analyses showed that the mutation destabilizes the conformation of the receptor, explaining why it affects its binding to both HGFF and HGFF kinase (followed by details of the changes). Xx mutations have been found to cause similar structural changes in several other receptors (followed by illustrative examples from the literature).
>
> Modeling of the interactions between the receptors, HGFF and the kinase, in the presence and absence of X189-X191, showed that these small molecules stabilize a non-native HGFF receptor conformation that presents active ligand-binding sites, thus explaining their enhancement of the receptor's affinities and further corroborating the tested hypotheses (with a summary of the relevant data). The modeling, and subsequent empirical assays, also indicated that five variants of X189-X191, designated XE1-XE5, may be substantially more potent than the lead compounds.

We should stress here, again, that we are not suggesting that this case study reflects any real interactions, we are merely using it to illustrate how sections of a complex study can be drafted.

> As also mentioned in the Introduction, animal models can be extremely valuable for testing the therapeutic potency of drug candidates (with illustrative references), therefore in addition to the affinity and structural tests we screened potential animal models by pp (essential reference) and identified Anon's rat (essential reference) as an apparently suitable model that both carries a corresponding mutation and displays Anon symptoms. Therefore, in the final part of the study, the toxicity of XE1-XE5 was tested by means of yy assays (essential reference), and their therapeutic potency was tested in trials with Anon rats. The results show that XE1 and 4 (but not XE2, 3 or 5) meet standard toxicity criteria for clinical trials, and substantially alleviated Anon symptoms in Anon rats.

4.2.7 Conclusion(s)

Finally, the concluding remarks could be based on the last of the summarizing statements:

> The results of this study corroborate the hypothesis that xx mutation in HGFF receptors can cause Anon's disease by affecting their affinities for HGFF and HGFFs kinase. They also confirm that X189-X191 increase the affinities of mutant receptors, strongly indicating that this is the mechanism whereby they slightly alleviate Anon symptoms, and that Anon rat is a suitable animal model for Anon's disease. In addition, two variants of X189-X191 more strongly enhance the affinities of the mutant receptors, meet toxicity criteria and (hence) may have considerable potential for treating the disease.

Thus, the described procedure can be used to generate clear, coherent drafts of all sections of even very complex papers, which can be conveniently used to identify essential and illustrative references that need to be cited (and where to cite them) and other information that needs to be incorporated when writing a complete paper.

4.2.8 Incorporated Sub-headings

Many journals prefer, or insist on, the use of sub-headings to divide various sections of papers into convenient sub-sections that are easier to read and understand than full, undivided sections. In addition to helping to compose sections of papers, brief summarizing statements can be conveniently used, with a little further abbreviation, to construct the sub-headings required in such cases. For example, the brief statements describing case study 4 could be used to construct sub-headings for a *Materials & Methods* section, if the target journal approves of their use, as follows

- *Expression of wild type and mutant forms of the receptor*
- *Determination of affinities of the receptors and complexes*
- *Structural characterization of the receptors and complexes*
- *Identification of an animal model*
- *Toxicity and therapeutic potency assays.*

With suitable modifications they can also be used to construct sub-headings for other sections, for example, the *Results*:

- *Receptors are expressed in correct conformation in ww*
- *Affinities of the receptors and complexes*
- *Structural features of the receptors and complexes*
- *Anon rat is a suitable animal model*
- *XE1 and 4 have therapeutic potency and meet toxicity criteria.*

Of course, if the target journal does *not* approve of their use, any sub-headings used to help structure a paper should be deleted before it is submitted.

Chapter 5
Linguistic Points

In this guide we have attempted to distil experience gained in many collective years of writing, editing and reviewing papers that we hope will help readers to maximize chances of their papers being accepted for publication. As we mentioned at the start, the main focus has not been on the grammar or construction of sentences in English (or any other language), but it is impossible to describe how to finalize drafts of papers without considering some linguistic aspects. Therefore, selected linguistic points are discussed in this section, starting with jargon and then: tenses, active and passive voices, practical considerations, unnecessary 'weak' verbs, narrative flow, plagiarism, and acceptable uses of other authors' works.

5.1 Jargon

The word jargon has three main meanings:

- phrases used by a particular trade, profession, or group, for example, medical jargon
- unintelligible or meaningless talk or writing, gibberish
- any talk or writing that one does not understand.

The meanings are connected, because jargon used by any professional group may not be understood by members of other professions. So, for instance, if your paper is on a biophysical process and you use too much biology and physics jargon, the physics may not be understood by biologists, the biology may not be understood by physicists and none of the paper may be understood by most people. Even the *Title* of your paper may be incomprehensible to everyone except a very few specialists.

For example, some papers have titles such as:

> UJARGON: a nutrient, vegetation and water-table based module for MIREMETH2 calculations

Such a paper may describe a module developed by staff of the (fictitious) University for the Promotion of Jargon for improving estimates of methane emissions from mires using the (fictitious) model MIREMETH2. The title (and paper)

J. Blackwell, J. Martin, *A Scientific Approach to Scientific Writing*,
DOI 10.1007/978-1-4419-9788-3_5, © Springer Science+Business Media, LLC 2011

could be made much easier to understand for non-specialists by mentioning this, by simply changing it to *UJARGON: a nutrient, vegetation and water-table-based module for estimating methane emissions from mires using MIREMETH2*. Further, we have frequently seen or edited papers in which key words, such as mires and methane, are not mentioned at all, because it is assumed (for instance) that readers will know that MIREMETH2 is intended for this purpose. This is a major mistake, for several reasons.

First, it restricts the potential readership to people who already know that this is the purpose of the model. Second, even if MIREMETH2 is currently widely used, in a few years it will probably be supplanted by MIREMETH3, or another model with a completely different acronym. Then, writers of reviews, or anyone else with an interest in the subject, will probably miss the study because they will probably use key words such as methane, mire and emissions in their literature searches. Hence, the number of possible citations will also be reduced.

For these reasons, jargon should be avoided whenever possible, and standard English words should be used instead, since its use reduces the number of people who will understand, read and cite a paper and annoys many who do understand it. Of course, there are many occasions when its use is unavoidable. For example, in combustion engine research a frequently used parameter is engine timing, expressed in terms of Crank Angle Degrees before Top Dead Centre or Crank Angle Degrees after Top Dead Centre. Clearly, it would be absurd to frequently repeat these phrases in a paper, so instead authors generally use the acronyms CAD bTDC and CAD aTDC, respectively. However, even in such cases it is preferable to mention the meaning of the terms when they are first used (or provide a list of acronyms). Similarly, numerous widely used scientific terms (e.g., expressed sequence tag, pulsar, bioinformatics and rheostat; four out of thousands of terms that could have been mentioned) could be regarded as jargon, but frequently have to be used, because otherwise several sentences (at least) would be required to describe the things they refer to. Nevertheless, as pointed out by George Orwell in his 1946 essay, *Politics and the English Language* (see Orwell 2010) whenever possible it is good practice to *Never use a foreign phrase, a scientific word, or a jargon word if you can think of an everyday English equivalent.*

5.2 Tenses

Tenses are often used inconsistently, and rather confusingly, even by experienced native-speaking scientists. Such inconsistency is exacerbated by over-use of the present tense, and often appears at the start of papers, when authors may state, for instance, *The aim of this study is to test the hypothesis that deer are deterred by wolf urine...*, *In this paper we explore relationships between the structures of a set of compounds (designated X1-200) and their effects on HGFFr activity...* or *We develop a model to...* This immediately leads to a problem, since the following text must either continue rather bizarrely in the present tense (e.g., ... *Deer are introduced into enclosures containing...* or *We analyze these compounds by GC-MS...*), or shift to the past tense (*Deer were introduced...*). Clearly, deer are not introduced into enclosures, no compounds are analyzed by GC-MS, and models are not developed in any paper. Instead, experimental procedures are applied in the field or laboratory, and models are developed in silico before a paper is written, as part of the underlying research. In addition, the *objectives* of a study are (obviously) formulated *before* any paper is written. Hence, such clashes in tense arise whenever parts of the *objectives* and *what was done* are expressed in the present tense. Fortunately, this problem can be avoided by simply using the past tense to describe elements of the *rationale* (apart from established truths, see later), *objectives* and *what was done*. For instance, the earlier examples could be simply changed to *The aim of this study was...*, *We have explored...* and *We have developed a model to...* Alternatively, the problem can sometimes be avoided by simply deleting unnecessary words, for example, *To test the hypothesis...*

The present tense is also often used inappropriately in modeling analyses to describe conditions that may arise in the future, for example, *We find that mean annual temperatures are xx °C warmer in Poland in 2100 than now.* Clearly, we will not know whether such a prediction is accurate for 90 years, and in the meantime the earth may have been destroyed by (say) a collision with a meteor. Hence, constructions such as *The simulations indicate that mean annual temperatures will be xx °C warmer in Poland in 2100 than now* are much better.

Another problem associated with over-use of the present tense is that authors often use it to describe *all* of their findings, without caveats, indicating that some findings that only apply at particular times or under highly specific conditions are currently (and generally) true. To illustrate the illogicality of using the present tense in such cases, a paper we edited reported estimated numbers of salmon in a certain river, stating that there *are xx thousand adult salmon in the river*, rather than that the authors' estimates indicated that there were xx thousand adult salmon in the river at the sampling time. Even if their estimate was accurate then, it is extremely unlikely to be accurate for other times because salmon are migratory, and they are subject to predation and mortality through other causes.

The tendency to abuse the present tense in this manner is encouraged in some cases by the *Instructions for authors* of various journals – for example, the *Journal of Bacteriology* (2010) and Nature Publishing Group's *free science library and personal learning tool* Scitable (Nature Education 2010) – *stipulate* that the present tense should be used to present authors' own conclusions, and conclusions of previous researchers, without always mentioning that this should only be done if suitable qualifications are added. Use of the present tense without appropriate caveats clearly implies that conclusions are currently (and generally) valid, hence its use in this manner implies that relationships that may have been observed under carefully controlled conditions (which may have been selected to ensure that desired results were obtained) are general truths. Subsequent authors may then cement this implication when citing the results. Indeed, *Instructions for authors*, and various commentators, often encourage such unwarranted extrapolation. For instance, Matthews and Matthews (2008) state that *When scientific information has been validly published in a primary journal, it becomes established knowledge.*

Thus, three major problems are associated with over-use of the present tense:

- It leads to statements that are almost certainly untrue (e.g., that xx thousand salmon *are* present in a certain river), or that may well prove to be untrue (e.g., that temperatures in a certain location are xx °C higher than current temperatures at some date in the future).
- It leads to unnecessary difficulties in describing sequences of events or investigations and clashes in tense.
- It indicates that some conclusions drawn from analyses of materials or processes under highly specific conditions are general truths.

These problems can be simply avoided by using: the past tense to describe the *rationale*, *objectives* and *what was done* in a study; the future tense to describe things that *may* happen in the future; and the present tense to describe well-established general *findings*, results that may only be applicable under certain conditions (with appropriate caveats) and the *implications* of results, again with appropriate caveats or conditional terms (e.g., *The results indicate that...*).

A further problem associated with tenses is that authors often ambiguously describe the results of previous studies using the simple past tense of to be (i.e., *was/were*), for example, *cancer levels were higher among smokers than among non-smokers (Hemingway 1999)* (fictitious reference). This too should be avoided because it is not clear whether it means that cancer levels used to be higher among smokers than among non-smokers (but may not be now), or cancer levels were higher in smokers than in non-smokers globally, or cancer levels were higher among the smokers than among the non-smokers in the cited author's set of subjects. Hence, it is far more informative to state that *cancer levels are higher among smokers than among non-smokers (Hemingway 1999)* if you believe that this is an established truth and are citing the reference to support the assertion. In contrast, *Hemingway (1999) found that cancer rates were higher among smokers than non-smokers amongst adults in the UK* is far more informative if you believe that the cited results applied to the subjects that Hemingway considered, but do not necessarily apply in other cases, especially if you are going to cite another study with conflicting results, for example, *In contrast, Bouchard and Flaubert (2000) found that cancer levels were higher among non-smokers in France* (strongly fictitious reference).

Non-native speakers (and many native speakers) understandably have many problems deciding what past tense to use, both when describing previous authors' findings and what they did in a study. However, there are a number of simple rules that can be followed to ease the decisions.

The simple past tense can be used, correctly, and unambiguously, when:

- Describing things that you did and the results you obtained in a presented study. Examples include, *Subjects and controls were matched according to age and gender* or *The bacteria were cultivated in shake flasks at 37°C* or *The plots were laid out in a fully randomized design.*
- Describing observations by other authors that only applied to certain materials or at certain times, and those materials or times are specified either explicitly

or implicitly. For instance, *Levels of cruciferoids were higher in the leaves of* Arabidopsis *seedlings than in the leaves of mature plants that Yan (2006) examined* or *Yan (2006) found levels of cruciferoids to be higher in leaves of* Arabidopsis *seedlings than in leaves of mature plants* (fictitious reference). Note, neither of these alternatives implies that the findings are necessarily always valid.

In contrast, when making assertions that you believe to be thoroughly established, generally valid facts, there are three possibilities:

- Use the present tense, accompanied by appropriate references, for example, *The causative agent of tetanus is the Gram-positive bacterium* Clostridium tetani *(reference)* or *Reactive oxygen species can be highly damaging to biological tissues (reference).*
- Use the present perfect progressive tense ('has been' or 'have been'), with references, in the active voice, for example, *Harrow (2006) has shown that levels of cruciferoids are influenced by temperature* (fictitious reference).
- Use the present perfect progressive tense ('has been' or 'have been'), in the passive voice, for example, *The causative agent of tetanus has been shown to be the Gram-positive bacterium* Clostridium tetani *(reference), It has been thoroughly established that reactive oxygen species can be highly damaging to biological tissues (references)* or *It has been postulated that xx mutation in HGFF receptors is a cause of Anon's disease (reference).*

Note, some extreme advocates of the active voice would greatly dislike some of the earlier examples because they are in the passive voice, which they think should rarely (or never) be used (e.g., Scarrow 2004). This is a mistake, because it prevents exploitation of much of the flexibility of the English language (possibly its greatest strength), as discussed in the following section (5.3) on active and passive voices (see Section 5.3).

It should also be noted that these are only selected points regarding use of tenses in English. A full description of their uses is beyond the scope of this guide, but there are many good books and web resources on the subject that can be used to obtain further information. However, there are also many very poor ones, especially those written by scientists, since few native English-speaking scientists have thorough grounding in grammar nowadays. Therefore, clearer general expositions can often be obtained from grammar books or dictionaries (in print or online), for example, *Merriam-Webster's Learner's Dictionary* (2010) and *The Encyclopædia Britannica* (2010). In addition, as always, the target journal's recommendations should be followed (unless their editors and papers published in them clearly do not follow those recommendations, see later).

5.3 Active and Passive Voices

Many journals' *Instructions for authors* include references to the use of the 'active voice' and 'passive voice'. The difference between these 'voices' can be illustrated by this short passage from The Encyclopædia Britannica (2010):

> English grammar distinguishes between the active voice ('The hunter killed the bear') and the passive voice ('The bear was killed by the hunter'). In the active voice, the emphasis is on the subject of the active verb (the agent performing the action named), whereas the passive voice indicates that the subject receives the action.

Other examples include *The wind disperses seeds* (active) versus *Seeds are dispersed by the wind* (passive), and *Temperature strongly affects rates of chemical reactions* (active) versus *Rates of chemical reactions are strongly affected by temperature* (passive). A related (but not synonymous) issue, considered later, is use of personal pronouns, for example, *I* (or *We*) *sowed the seeds* (personal, active) versus *The seeds were sown* (impersonal, passive).

During much of the twentieth century, the passive voice was preferred in science writing since it was considered to be more objective (e.g., if non-dormant seeds are placed in an appropriate growth medium under appropriate conditions, they should germinate regardless of who places them in the medium). Furthermore, most readers do not care who planted seeds or analyzed extracts, they merely want to know how things were done and the results. For these, and other, reasons, use of the active voice was often strongly discouraged in almost all circumstances.

However, in recent decades there has been a strong swing toward use of the active voice. Indeed, many commentators and style guides state that the active voice should generally, or even *always*, be used (e.g., Young 2006). In an article entitled *Common errors in technical writing*, John Owens even approvingly cites a quotation from Simon Crowley (unidentified original source, probably a Twitter 'tweet') stating that *Every time you use the passive voice, a kitten is killed by God* (Owens 2010). Numerous journals also prefer its use, including *Nature* (Nature 2010a) and *Science* (Science Magazine 2010).

There are several reasons for the rising preference for the active voice. First, it advocates claim that its use can make writing more lively, more succinct and avoid awkward phrasing. There is some truth in this. For instance, *The eagle caught the fish* may, in many circumstances, be better than *The fish was caught by the eagle*. Often, it is also less ambiguous, for example, if a paper states that *It was concluded that smoking is a major cause of lung cancer*, there are no indications who reached the conclusion. Thus, if the authors mean they reached it, they can more informatively state *We conclude that smoking is a major cause of lung cancer* or *Our results suggest that smoking is a major cause of lung cancer* or *Therefore, smoking appears to be a major cause of lung cancer.*

However, the passive voice has many valid uses. Many things can be stated much more succinctly in the passive voice, for example, *English is widely used* or *Rice is sold by the kilo* would require much more awkward constructions in the active voice. In addition, many of the most famous, elegant and moving passages in English are in passive voice, partially or entirely, for instance

> We hold these truths to be self-evident, that all men are created equal... (United States Declaration of Independence 1776).

and

. . .one hundred years later, the life of the Negro is still sadly crippled by the manacles of segregation and the chains of discrimination (Martin Luther King *I have a dream* speech 1963).

and

Never in the field of human conflict was so much owed by so many to so few (Winston Churchill 1940).

and

To be or not to be – that is the question:
Whether 'tis nobler in the mind to suffer
The slings and arrows of outrageous fortune,
Or to take arms against a sea of troubles
And, by opposing, end them. To die, to sleep
No more – and by a sleep to say we end
The heartache and the thousand natural shocks
That flesh is heir to – 'tis a consummation
Devoutly to be wished. To die, to sleep
To sleep, perchance to dream.
(Hamlet, William Shakespeare ca. 1600).

Thus, although there are good grounds for *generally* using the active voice, the passive has important uses (Merriam-Webster 2010), notably:

- when the receiver of the action is more important than the individual undertaking it, for example, *The leaves were sampled*
- when the doer is unknown, for example, *The third deer was only tracked for two days because it was killed by a predator during the second night*
- when the speaker or writer wants the individual responsible for the action to remain anonymous, for example, *Mistakes have been made* (a useful construction in reports, when you do not want to name the person who made the mistakes because it was you, a person you want to protect, or someone who could end your career).

It should also be noted that, in evolutionary terms, the passive voice evolved because there were strong selective pressures; in various situations it is much better than the active voice. Furthermore, it is still usually used, even in journals that state a preference for the active voice, in *Materials & Methods* sections, because (as mentioned earlier) for nearly all readers *what was done* is of *much* more interest than who did it.

Hence, nearly all journals would prefer *The seeds were germinated on moist filter paper* rather than *We germinated seeds on moist filter paper.* To illustrate the strength of this preference, Google returned 6,520 hits for *We sowed the seeds* and 213,000 hits for *The seeds were sown* (9 August 2010). Since the message is usually more important than the messenger, persistent use of the active voice when describing previous authors' results can also lead to very stilted writing that stresses the cited authors rather than their findings, for example:

> Young (1989) first described cruciferoids in the late 1980s. Yan (1993), Hornberg (1994) and James and Wang (1994) later found indications that they are involved in responses of certain plants to water stress, since their levels rose strongly in shoots during drought periods. However, Carlsson (1995) and Gonzalez (1996) found contrary indications in several cases, casting doubt on the hypothesis. Strindberg (1998) and Harrow (2000) subsequently found that cruciferoids' effects are modulated by light and temperature, respectively, providing possible explanations for the apparently conflicting results (all fictitious references).

The passive voice can be much clearer (and just as succinct) for such passages, because it places the emphasis on the *topic of interest* (cruciferoids) rather than the *cited investigators*, for example:

> Cruciferoids were first described in the late 1980s (Young 1989). Indications that they are involved in responses of certain plants to water stress were subsequently found, notably that their levels rose strongly in shoots during drought periods (Yan 1993, Hornberg 1994, James and Wang 1994). However, contrary indications were found in several cases (Carlsson 1995, Gonzalez 1996), casting doubt on the hypothesis. Later findings that their effects are modulated by light (Strindberg 1998) and temperature (Harrow 2000) provided possible explanations for the apparently conflicting findings.

This is especially true when a numerical reference system is being used, since then active voice constructions require absurdities such as:

> Authors of reference [1] first described cruciferoids in the late 1980s. Authors of [2,3,4] later found... or [1] first described cruciferoids in the late 1980s. [2,3,4] later found...

A related grammatical issue, as mentioned before, is whether or not to use personal pronouns (e.g., *I/We sowed the seeds* versus *The seeds were sown*). There has been a shift in preferences in recent decades, from impersonal to personal, that is similar (but weaker) to the shift from passive to active. The main justifications for this are that saying *The seeds were sown* or *It was concluded* is less informative than personal forms (*I sowed the seeds* or *We concluded*), as previously mentioned, and failing to state that *I/We did it* is false modesty or a 'conceit' (elaborate pretence, i.e., you know you did it, your readers know you did it, so why not say so?).

However, use of the personal voice is also often a conceit. For example, *We sampled leaves from trees in the greenhouse* implies that all of the authors trooped into the greenhouse and collected leaves, when probably just one did, or just as likely a technician working in the greenhouse (say Mrs Harris) took them on every sampling occasion, except one Tuesday when she was ill so someone else collected them. Strictly, this should be described in active voice as follows, *Mrs Harris (a technician working in our institute's greenhouse) sampled leaves on every sampling occasion, except one Tuesday when she was ill so someone else (we cannot remember who now) sampled them.* In practice, the key message is that the leaves were sampled, so use of the passive voice (*Leaves were sampled from plants in the greenhouse*) is much clearer, more concise and thus better.

Furthermore, overuse of *I/We* becomes annoying. For instance, if there are long passages stating that *We grew the plants... We then sampled... We extracted the leaves... We analyzed the extracts by GC-MS... Our results show... We conclude...* the authors start to seem like bores at a party who will only talk about themselves, since the emphasis in all the sentences (the 'subject') is on them, not *what was done* and the *findings*. The impression given is that the main intention of the authors is to show how diligent and clever they were. Of course, one of the main intentions of the authors of *any* paper is to show they were clever – but there are more subtle ways to do it. Therefore, more sparing use of *I/We* is generally better, interspersing sentences written in the active, personal voice with others written in passive voice or active but impersonal voice (e.g., *The plants were grown* or *The GC-MS analysis showed...*).

Particular care should be applied when stating *We conclude*, because it invites some readers (including referees and editors of target journals) to respond by thinking *Well, you might conclude that, but I'll make up my own mind* or (worse) *Rubbish, the results don't show that at all.* Thus, it is often better to state *The results show that...*, *The data suggest...* or, simply, *Therefore...* (depending on how definitive the evidence is, and the preferences of the target journal).

It should also be noted that many well-qualified editors and authors are confused about the distinctions between the active/passive voice and the impersonal/personal voice. For example, a member of the *Nature* editorial team states, in answer to a query, that ... *the active voice ('I' or 'We' in the case of multi-author papers) is better and makes papers far clearer and more comprehesible* [sic] (Clarke 2007). She goes on to say*: Active voice has been* Nature *policy for as long as I can remember; it is enshrined in our style manual and is specifically recommended to all authors as part of our standard acceptance procedure* (Clarke 2007).

Note, the editor's response is partly in the passive voice, and uses religious terminology (*it is enshrined and is specifically recommended*). Furthermore, the first sentence of a summary used to illustrate *How to construct a* Nature *summary* on the journal's website is in passive voice: *During cell division, mitotic spindles are assembled by microtubule-based motor proteins* (Nature 2010b).

In addition, in a randomly chosen volume of *Nature* (dated September 16, 2010):

- The Editor's summary begins with three sentences written partly or entirely in passive voice: *Induced pluripotent stem (iPS) cells are produced by reprogramming differentiated adult cells using a cocktail of transcription factors. They share many properties that are characteristic of embryonic stem (ES) cells generated by somatic-cell nuclear transfer (SCNT), and of ES cells from naturally fertilized embryos. The three cell types are not identical, however, and an interesting difference has now been discovered....*
- Some sentences in the summarizing paragraphs of all of the articles are partially or entirely in the passive voice.
- The *Materials & Methods* sections of all the articles are written largely or entirely in passive voice.

Thus, the meanings of active and passive voice are widely misunderstood, and the 'rules' concerning their use are often not followed, even by staunch advocates (because they are preferences, not rules, and they are not always applicable).

5.3.1 Practical Considerations

There are strong grounds for generally using the active voice, but like most general rules it is not always valid. The passive voice is often preferable when:

- You are stating what you did.
- Discussing the results of previous studies that have not been confirmed to be universally applicable; for example, *Concentrations of cruciferoids have been found to be higher in shoots of certain plants under water stress than in unstressed counterparts (Yan 1993), but their levels may also be strongly modulated by light (Strindberg 1998) and temperature (Harrow 2000)* (all fictitious references).
- You can state something more clearly and concisely in the passive voice. Clarity, as always, should be paramount.

In addition:

- Journals' instructions regarding active and passive voices should be followed, except in cases where their editors and the authors of papers published in them clearly do not follow stated preferences.
- Distinguishing between active, passive, personal and impersonal voices is not always easy, even for well-qualified native speakers. If you are not sure about the distinction, do not worry about it and concentrate on writing as clearly as possible (it is much more important, and if necessary the 'voice' can be altered later).

5.4 Unnecessary 'Weak' Verbs

Another related issue is that authors often use 'weak' (i.e., unnecessary or uninformative) verbs when describing what they did. For example, they may state that *Analysis of the compounds was then performed by nuclear magnetic resonance spectrometry* or *Sampling was then conducted by harvesting leaves*, rather than simply *The compounds were then analyzed by nuclear magnetic resonance spectrometry* or *Leaves were then sampled*, respectively. If you realize that you have used such unnecessary verbs in a paper, they should ideally be deleted and the sentences containing them should be modified so that they include only the 'stronger', that is, more informative, verb.

5.5 Narrative Flow and Coherent Arguments

In order to write a paper (or any communication) clearly, and present coherent arguments, it is essential to have a clear idea of the information you want to convey at every level, from the overall paper down to single sentences. Otherwise, there will inevitably be many 'jumps' from one disconnected idea to another that disrupt the narrative flow and hinder readers' ability to grasp the ideas and information you want to convey. Indeed, it will be impossible for readers to follow any clearly structured arguments, simply because there is no clear structure.

It is not always easy to structure the presentation of ideas and information when complex concepts or investigations are being described, but it can be greatly facilitated by following a systematic approach for structuring papers, such as that presented in this guide. Indeed, the key aim when drafting a paper systematically, as detailed for case studies 1–4, is to obtain such a structure, which indicates precisely the information that needs to be incorporated at each point in each section. Thus, a systematic draft will already have provided much of the structure needed to finalize a paper, but there is still a little more work to be done. Let us now review how much structure it provides, and the additional structuring that is required, at every level.

5.5.1 The Overall Paper

In order to maintain narrative flow at this level, it is obviously essential to present the information regarding the main aspects of a study (the *rationale*, *objectives*, *what was done*, the *findings* and their *implications*) in a clear, logical order. Therefore a paper must have a coherent overall structure. Fortunately, as discussed earlier, the standard format for scientific papers – *Title, Abstract, Introduction, Materials & Methods, Results, Discussion, Conclusion(s)* and *References* – provides a well-established template. In addition, journals' *Instructions for Authors* usually provide sufficient information regarding sub-titles that should be used for each of these sections and the order in which they should be presented (e.g., the *Abstract* may be called the *Summary*, and it may be presented after the *Title*, or at the end of the main text). Hence, we do not have to worry about structuring at this level, provided the journals' instructions are followed and we are aware of the following:

- *Title*: Should encapsulate the main *implications* or *findings* of the study (or *objectives* if the *findings* are too complex to outline in a title)
- *Abstract*: Should briefly describe the *rationale*, *objectives*, *what was done*, the *findings* and their *implications*
- *Introduction*: Should describe the *rationale* in detail, then briefly summarize the *objectives* and *what was done*
- *Materials & Methods*: Should describe in detail *what was done*
- *Results*: Should describe in detail the *findings*
- *Discussion*: Should briefly recapitulate the *rationale*, then describe in detail the *implications* of the results
- *Conclusion(s)*: Should briefly recapitulate the *objectives*, *findings* and major *implications* of the study.

5.5.2 Sections of Papers

We have already considered in detail the way in which knowledge of the information that needs to be presented in each section of a paper (in conjunction with brief statements regarding the *rationale*, *objectives*, *what was done*, *findings* and the *implications* of a study) can be used to write a *Title* and complete *Abstract*. Thus, structuring and maintaining narrative flow in these sections will not be further considered here.

Maintaining narrative flow in other sections of papers requires a little more thought, because they are much longer, and journals' *Instructions for Authors* usually provide little or no advice on their structure. However, as already discussed at length, drafts composed using appropriate sub-headings (which may or may not be included in the final text) based on brief statements regarding the *rationale* and so on can provide a framework that highlights the information and references that need to be incorporated (and the reason they need to be incorporated) at each point in every section. Thus, such drafts provide a scaffold of a structured flow of information. All that remains to be done is to insert, and coherently link, the information and references.

We have also considered the tenses that should be used (e.g., that the past tense should generally be used to describe things you have done), the 'voice' that should be used (generally active, except when information can be more clearly or concisely stated in passive voice) and the desirability of avoiding jargon, when possible. A further linguistic aspect that needs to be considered, before writing final versions of sections of paper, is the use of conjunctions (words or phrases that connect words, phrases, clauses or sentences) to link paragraphs and sentences.

5.5.3 Paragraphs and Sentences

There are huge numbers of conjunctions (e.g., and, furthermore, but, or, not, either, after, therefore, because, since, as, if, when, for example and for instance). It would be impossible to describe and illustrate uses of all the conjunctions that could be used in a short book, and plenty of dictionaries are available that comprehensively detail their uses. Thus, we do not focus on any particular words here. Instead, we simply note that choosing appropriate conjunctions to link ideas and information is essential, and if you are unsure about the best choice in a given situation, a dictionary should be consulted.

It is also impossible to illustrate all the kinds of arguments or logic that may need to be deployed, since they will be different in every paper. However, as an example, if you want to cite several studies that support an assertion you could write:

> Numerous authors, *inter alia* Yan (1993), Hornberg (1994) and James and Wang (1994) have found that cruciferoids are involved in responses of certain plants to water stress (all fictitious references).

If you then want to contrast these findings with conflicting results, you could modify this sentence and add:

> Cruciferoids were first described in the late 1980s (Young 1989). Indications that they are involved in responses of certain plants to water stress were subsequently found, notably that their levels rose strongly in shoots during drought periods (Yan 1993, Hornberg 1994, James and Wang 1994). However, contrary indications were found in several cases (Carlsson

1995, Gonzalez 1996), casting doubt on the hypothesis. Later findings that their effects are modulated by light (Strindberg 1998) and temperature (Harrow 2000) provided possible explanations for the apparently conflicting findings.

While composing, and linking, clauses, sentences or paragraphs, it is important to avoid non sequiturs (in this context statements that do not follow logically from preceding statements). Authors frequently introduce non sequiturs because they have knowledge that they assume readers will also have, for instance, in a paper describing an investigation of effects of over-fertilization of certain crops in a certain area, they may state that *Since over-fertilization can cause environmental problems, we assessed the possibility that surface waters in the area are eutrophic*. This is a non sequitur because it fails to add a key piece of information; that one of the environmental problems that over-fertilization can cause is eutrophication. Therefore, this passage should be amended, as follows: *Since over-fertilization can cause environmental problems, notably eutrophication (refs.), we assessed the possibility that surface waters in the area are eutrophic.*

Of course, in many paragraphs, especially in the *Introduction* and *Discussion*, findings and conclusions of other authors (e.g., that over-fertilization can cause eutrophication) need to be summarized. The only sources of information about other workers' results and conclusions are the primary sources (their papers), and/or secondary sources (e.g., reviews) that describe them. However, the relevant information provided by these sources must be presented without plagiarism, therefore, this is the last linguistic aspect that will be explicitly considered here.

5.6 Plagiarism and Acceptable Uses of Other Authors' Works

There are two kinds of plagiarism that journals and referees will not accept (if detected): the presentation of other authors' conclusions without attribution, that is, without reference to the works in which the conclusions were first stated and word-for-word quotation of passages from other authors' papers, unless the quoted words are enclosed in quotation marks or italicized and duly referenced. Both of these kinds of plagiarism are regarded as breaches of copyright conventions and hence, essentially, theft of intellectual property.

However, viewing the latter type as plagiarism can be regarded as another 'conceit', in a sense, since as mentioned earlier, other authors' works are the only sources of information about their findings and conclusions. Thus, their words (and/or mathematical formulae and data) have to be used when constructing paragraphs describing what they did, found and/or concluded, but with suitable *paraphrasing*, that is, expressed in other words. This is also often essential for the sake of concision, since a whole paper may be devoted to showing, for instance, that levels of cruciferoids are modulated by temperature, but in a particular paper you may only want to state *Levels of cruciferoids are also modulated by temperature (Harrow 2006).*

Having considered the linguistic points outlined earlier, we are almost ready to finalize papers, using drafts constructed as illustrated earlier showing the information and references that each section must present, the order in which they must presented be and the messages that need to be conveyed, using appropriate conjunctions. Presenting specific examples here (e.g., *Introductions* or *Discussions* for the case studies) would be inappropriate, and potentially misleading, because their length, the amount of detail that should be included and the linguistic style would all depend on the target journal's preferences. To see examples of how this is done, we recommend reading as many as possible articles regarding your fields of interest to examine how other authors have structured the presented information, in addition to accessing the information per se.

References

Clarke M (2007) Active is better than passive or neutral. Nautilus. http://blogs.nature.com/nautilus/2007/01/active_is_better_than_passive.html. Accessed 10 Aug 2010

Journal of Bacteriology (2010) Instructions to authors. http://jb.asm.org/misc/journal-ita_abb.dtl#01. Accessed 20 Sept 2010

Matthews JR, Matthews RW (2008) Successful scientific writing: a step-by-step guide for the biological and medical sciences. Cambridge University Press, New York, NY

Merriam-Webster (2010) Learner's dictionary: grammar glossary. http://www.learnersdictionary.com/help/glossary2.htm. Accessed 10 Aug 2010

Nature (2010a) How to write a scientific paper. http://www.nature.com/authors/author_services/how_write.html. Accessed 1 Dec 2010

Nature (2010b) How to construct a Nature summary paragraph. Information sheet 2c. http://www.nature.com/nature/authors/gta/2c_Summary_para.pdf. Accessed 5 Sept 2010

Nature Education (2010) Effective writing. http://www.nature.com/scitable/topicpage/effective-writing-13815989. Accessed 20 Sept 2010

Orwell G (2010) Politics and the English language. In Politics and the English Language and other essays. Benediction Classics, Oxford, UK

Owens J (2010) Common errors in technical writing http://www.ece.ucdavis.edu/~jowens/commonerrors.html. Accessed 1 Sept 2010

Scarrow R (2004) Haverford Chemistry Department general guidelines for written reports www.haverford.edu/chemistry/resources/writingguidelines.pdf. Accessed 1 Dec 2010

Science Magazine (2010) Some notes on *Science* style. http://www.sciencemag.org/about/authors/prep/res/style.dtl. Accessed 10 Aug 2010

The Encyclopædia Britannica (2010) Voice. http://www.britannica.com/EBchecked/topic/631842/voice. Accessed 9 Aug 2010

Young P (2006) Writing and presenting in English: the Rosetta Stone of science. Elsevier, Amsterdam, The Netherlands

Chapter 6
Covering Letters and Referees' Objections

Having written a paper to the best of your ability, the next step is to submit the paper, which should be accompanied by a carefully composed covering letter, to the editor of the target journal. Then, the final obstacles that usually have to be overcome are referees' objections. This is part of a process designed to improve papers, and ensure that only those of sufficient quality are published by journals, thus safeguarding their reputation and hopefully avoiding the dissemination of poor or sloppy work. The process is far from perfect, and it should be recognized from the outset that however carefully you have designed and executed your experiments, and however brilliant the study may be, your paper may still be rejected or changes may be required, because of deficiencies (real or imagined) noted by referees. To maximize the chances of a paper being accepted, it is essential to be as aware as possible of potential objections long before submitting it and to adopt appropriate strategies for countering any objections that may be raised, whether they have been anticipated or not. Composing a covering letter, anticipating referees' objections and countering them are the foci of this chapter.

6.1 The Covering Letter

The covering letter, the letter accompanying a paper submitted to a journal, is highly important, since it tells the editor whether or not the paper is likely to be of interest to the journal's readers before he/she, or any referee, reads it. The editor may, or may not, read the paper before sending it to referees and may, or may not, add a few comments regarding initial opinions about the paper when it is sent to them. If the editor does add comments, it is clearly highly desirable that they should be something like, *We have received the attached paper, which looks very interesting from the covering letter...*, rather than *We have received the attached paper, which was accompanied by an incomprehensible covering letter, suggesting that the paper is likely to be pretty incomprehensible too....* Obviously, the contents of the letter will depend on the subject of the study. However, in essence they should always be the same since they should *politely* state the *Title* of the paper, and outline (as in an *Abstract*) the *rationale*, *objectives*, *what was done*, the *findings* and the *implications*

J. Blackwell, J. Martin, *A Scientific Approach to Scientific Writing*,
DOI 10.1007/978-1-4419-9788-3_6,

of the study, except in a little less detail, since brevity is even more important than in an *Abstract*. It is also essential to highlight novel aspects of the study (either implicitly or explicitly), as it is in the paper itself. Thus, the ideal covering letter should provide sufficient information for a busy editor to scan very quickly and leave him or her thinking *Interesting study!* so he or she can pass this verdict to the referees. Of course, the referees may disagree, having read the paper in more detail, but this does at least give the best possible start to the process.

Any covering letter should begin with *Dear Sir/Madam*, if the name of the editor is not known, or *Dear Prof. Li* (if, for instance, it is known that this is the name of the editor). To illustrate the rest of the contents of such a letter, examples for case studies 1 and 2 (based on the brief statements regarding their *rationale*, *objectives*, *what was done*, the *findings* and *implications*) are presented later.

6.1.1 Hypothetical Case Study 1

I am (or *We are*, as appropriate) submitting a paper entitled '*Urine of wolf subspecies* imaginary *deters* Cervus unreal *deer from grazing crops in northern Sucrosia*' for consideration in *Crop defense* (fictitious journal).

Substantial proportions of crops in northern Sucrosia are lost to grazing by deer (*Cervus unreal*) roaming from neighboring hills. Current methods for deterring the deer are inhumane, expensive and/or ineffective. Applications of volatiles from urine of wolves could be cheaper and more effective, but first it is necessary to determine whether wolf urine is an effective deterrent. To test this possibility we applied urine from wolves (subspecies *imaginary*) around plots of wheat, maize and pea crops surrounded by grass meadow in enclosures, then introduced deer into the enclosures. The deer completely avoided these plots for two days, but rapidly consumed crops in control plots for two days. The results indicate that urine of wolf subspecies *imaginary* is a potent deer deterrent.

We believe that the findings will be of great interest to farmers in northern Sucrosia, and potentially to agriculturalists in other areas facing similar problems, thus we believe it warrants publication in *Crop defense*. We hope you will agree, and we look forward to your response.

6.1.2 Hypothetical Case Study 2

I am (or *We are*, as appropriate) submitting a paper entitled '*Coverage of fictional knotweed, stand density and alleles* res2 *and* res4 *strongly affect damage in* Pinus fictitious *populations by* Hylobius fabricated' for consideration by *Tree-insect interactions* (fictitious journal).

A previously minor insect pest (the weevil *Hylobius fabricated*) of young pine trees has begun to attack, aggressively, some (but not all) populations of *Pinus fictitious*, a threatened species of pine in northern Sucrosia. Both environmental and genetic factors are likely to affect levels of damage in these populations. Furthermore, the invasive weed fictional knotweed has spread extensively, and may be greatly affecting interactions between plants

> and insects in the region. We have examined three lightly damaged and three heavily damaged populations of the pine to test these hypotheses. Our results show that, as expected, both environmental factors (especially coverage of fictional knotweed and stand density) and genetic factors (alleles at quantitative trait loci designated *res2* and *res4*) strongly and interactively affect levels of damage by the pest.
>
> These findings provide clear indications of the measures that should be taken to protect the remaining populations of the pine, and/or replace destroyed forests, which we think will be of considerable interest to your readers, thus we hope you will consider our manuscript favorably, and we look forward to your response.

As usual, for any letter to anyone other than a close friend, colleague or family member, the letter should then end with Yours Sincerely, Faithfully or Truly, followed by the name of the corresponding author.

6.2 The Review Process

It has been widely reported on the web and elsewhere that Hans Krebs' seminal work on the biochemical sequence of reactions that now often bears his name (the Krebs cycle, otherwise known as the Tricarboxylic Acid or TCA cycle) was rejected by *Nature*, because it was deemed at the time to be of insufficient scientific merit for publication in the journal. This is not strictly true, the journal told him that it already had sufficient 'letters' for some time, and the editor would be prepared to keep his paper in the hope of making use of it when the 'congestion' cleared. Nevertheless, the response (and various cases of initial rejection of high-profile papers that were subsequently published in either the same or some other journal) shows that it is never certain that a paper will be accepted immediately, regardless of the brilliance of the study. This is partly because the attitudes of referees are unpredictable, and they may reject papers for diverse reasons, some of which have little to do with pure scientific appraisal. However, referees also often raise perfectly valid objections, and their comments can greatly improve papers. Thus, to maximize chances of publication it is essential to anticipate their objections (if possible), and to address them carefully. Strategies for doing this are discussed in detail later, but first the review process should be outlined.

The traditional review process can be summarized as follows:

- The editor of a journal sends each paper received to two or three anonymous 'peers', that is, experts in the subject covered by the respective papers.
- The referees read the papers and recommend that they should be accepted with no change (a rare event), rejected or revised (usually providing detailed lists of recommended changes).
- The editor decides whether to reject each paper or accept it, with no changes, minor amendments or major revisions.
- He/she then informs the authors of the decision, usually providing a list of referees' comments, and his/her comments, that they should address.

- The authors address the comments and return a revised version of the paper to the editor, with a covering letter detailing responses to the comments and amendments they have made (unless they decide to submit the paper elsewhere, or conclude that it is not worth re-submitting it anywhere).
- The editor considers the responses, and if he/she decides that concerns have been dealt with adequately, he/she may accept the paper without further review.
- Alternatively, he/she may return the revised paper and authors' responses to the referees, asking them if their concerns have been addressed satisfactorily and if the revised paper is of sufficient quality for publication.
- In some cases, there may even be three rounds of review and response.

The traditional review process is still applied by most journals. However, it has been heavily criticized by many commentators (e.g., Bishop 1984) since it is slow and the anonymity of the process may conceal personal biases, unethical behavior or incompetence. An obvious flaw is that the most highly qualified reviewers for evaluating a study are likely to be competitors of the authors (Relman and Angell 1989). Thus, for instance, there is a risk that referees may be unduly harsh because acceptance of a paper may prevent publication of data that they have been gathering.

Various editors and other commentators have defended the traditional system, for example, Gurr (1990). However, in addition to the published criticisms of this system, most experienced authors have some anecdotal evidence that referees' objections are not always entirely objective, and that referees' personal motives, perceptions and prejudices may affect their judgments of papers. For example, a paper written by one of us was initially rejected because it was supposedly not on an appropriate subject for the target journal. However, we knew, and were able to tell the editor of the journal concerned, that the referee had published several papers on the subject in the same journal. It was then accepted, and became a 'citation classic'. The only explanation for the referee's objection we can think of was that he or she was annoyed not to have thought of the described procedure (which was very simple).

Due to such concerns some journals have adopted or trialed 'open review' processes. For instance, in June 2006, *Nature* launched a parallel open peer-review experiment, in which some articles that had been submitted to the journal's regular, anonymous process were also available online for open, identified public comment. Open review systems remove anonymity from referees and publicly display any bias in their comments. However, they also have disadvantages, notably poor papers are not filtered out before presentation to the research community, and papers are seen in a raw state, in which there may be numerous embarrassing errors that would probably have been removed in the traditional, anonymous review system before publication. Various formats of open review have been applied, but they will not be further described here because, in practice, it does not greatly matter whether your paper is submitted to open or anonymous review; the required responses in both cases are very similar. The rest of this section is devoted to addressing referees' objections. This should be done:

- chronologically
- systematically
- calmly
- courteously.

6.3 Anticipating Objections

The process is chronological because it should begin *long before sending a paper*, by anticipating objections while planning and executing any study, and while writing any paper.

6.3.1 Anticipating Objections While Planning a Study

This is the *most* important stage for dealing with referees' objections. It is essential to:

- select appropriate materials (e.g., sites, populations or tissues) and controls to examine for your purpose
- apply appropriate treatments and/or sampling strategies
- measure all the key variables using robust methodology
- record all observations carefully.

Often, if (say) important controls are not included at the outset it will be impossible to incorporate them at later stages. Thus, papers describing the study may be doomed, long before they are even written, so it is critical to ensure that the experimental design is as rigorous as possible, or at least adequate to meet a strict referee's requirements.

Therefore, having planned a study, the plans should be considered as critically as a hostile referee would do, because eventually, when papers describing it are submitted, the experimental design will be intensively scrutinized. Furthermore, it is easier to spot flaws in other people's plans than in one's own plans. Therefore, having planned a study we recommend showing the detailed plans to experienced colleagues to check that you have not omitted anything important.

6.3.2 Anticipating Objections While Executing a Study

It is equally important to consider possible objections while actually performing a study, for the same reasons, that is, if experiments are executed in a sloppy manner, proper controls are not examined and/or full notes are not kept, it may be impossible to publish your results at a later stage. In addition, materials, samples and reagents used in the study should be kept after writing and submitting the paper, if possible, since referees may require additional experiments to be performed, and it may be possible to perform them quite quickly if they have been retained, but not otherwise.

6.3.3 Anticipating Objections While Preparing and Writing a Paper

Nature's response to Hans Krebs' work shows that a study may be rejected, or its publication may be delayed, regardless of how significant and novel it is. Indeed, if it is sufficiently revolutionary, it may cause wider social and cultural turmoil (as shown by responses to the work of Darwin and Galileo, see Section 3.7). It is not always possible to predict how referees may react to a paper, but it is important to be aware of objections that referees *may* raise, and prepare arguments to counter them as far as possible while preparing to write a paper. For this, presenting your study at a seminar or conference can often be valuable, since the responses to it may provide good indications of possible referees' objections, and highlight potentially problematic aspects that should be addressed before submission. Hence, opportunities should be taken to present it at appropriate events where possible (provided this does not alert other groups researching the same phenomena to the probability that you will be shortly publishing, and thus prompt them to submit similar findings before you).

Possible objections should also be carefully considered, of course, while writing the paper. Usually, referees' objections are based on one or more of the following:

- lack of relevance, novelty or limitations of the study
- concerns, or misunderstandings, about what you have done, how you have done it, or why you have done it
- linguistic concerns.

In order to avoid, or at least minimize, objections regarding lack of relevance, novelty or limitations, it is essential to highlight your study's novelties, and justify its *focus* in the paper, as discussed in detail in preceding sections. It is usually perfectly obvious to authors why they selected the systems, sampling sites/materials and methods they used, but it will not be so apparent to referees unless the reasons are clearly presented. Similarly, the *rationale* should be clearly outlined, to emphasize the importance and relevance of the subject. Authors are often leading experts in their fields, so concepts that seem obvious to them may not be at all obvious to other readers. Hence, it is essential to avoid omitting key information that general readers will require to follow your arguments, and to check that there are no non-sequiturs (see Section 5.5.3).

It is also sensible to minimize potential problems by tactfully describing results, and their implications, that conflict with hypotheses favored by likely referees. It is unwise to state that *Our results clearly show that the hypothesis proposed by Smith (2005) is incorrect* (fictitious reference), if Smith or someone who strongly supports his/her hypothesis is likely to be a referee (unless you know that Smith is one of those rare, good scientists who welcome being proved wrong). Instead, it is better to be more diplomatic, and perhaps state *Our results indicate that aspects of the hypothesis proposed by Smith (2005) should be re-evaluated, since it does not seem to be fully consistent with the observed phenomena, suggesting that other factors may be involved.* Better still, depersonalize the comment, perhaps as follows, *Our results indicate that factors other than water stress may influence cruciferoid contents.* Note, this may even be another factor to consider when selecting a target journal, since if your paper casts serious doubt on a hypothesis strongly advocated by someone who is likely to be a referee when you send it to a particular journal, it may be better to submit elsewhere.

6.3.4 Anticipating Objections After Submitting a Paper

Regardless of how carefully studies are designed, performed, tested in arenas such as conferences and presented, they are very rarely accepted with no demands for revision. Therefore, as many as possible of the objections that may still be raised need to be anticipated after submitting a paper, because although a journal may often keep a paper for a long time before telling authors whether or not it will be accepted, they often require responses to referees' objections to be submitted very quickly.

For example, sometime between submitting a paper and receiving a decision from the journal's editor, authors may think of something they should have done (e.g., controls they should have included). In such cases, if time and resources allow, it is wise to start performing omitted tasks before receiving the editor's decision. Since journals often take several weeks (or more) to consider papers, it may be possible to complete them before the editor informs you of his/her decision. If so, some of the referees' objections may have already been addressed. It may even be possible to counter humiliating comments such as

> The authors have made elementary mistakes, including the omission of critical controls and failure to determine levels of the analytes in key variants.

by stating

> The referee is quite right. We did include these controls and measure the analytes in the mentioned variants, but due to an oversight we did not mention them in the paper. We apologise for this error, but we have incorporated the relevant information in the revised paper (then stating the pages and lines where the information has been inserted).

Similarly, it may be possible to perform additional experiments that you have not thought of but the referees require quite quickly (especially, if materials, samples and reagents used in the study have been kept). You can then state:

> The referee is right that these experiments should have been included, and they were. We did not incorporate the results previously, but we have included them in the revised paper (again stating the pages and lines where the information has been inserted).

In addition, before receiving the editor's decision, you may think of points that should have been added about the *rationale* or *implications* of the study. If so, they should be jotted down. Then, if they are the basis of referees' objections, you can write:

> These are clearly points that should have been included in the Introduction/Discussion (as appropriate), so we have added them to the revised document (again stating the pages and lines where the points have been inserted).

Finally, it is *crucial* to be aware that

- very few papers are accepted with no demands for revisions
- time will probably have to be set aside to deal with referees' objections
- referees will very likely see flaws in your study (real or imaginary) that you have missed.

In addition, their comments may be extremely hostile or rude, although most journals provide clear guidance to their referees to avoid such comments (see, for instance, the detailed description of the *British Journal of Nutrition*'s procedures presented by Gurr 1990). A few real examples of unnecessarily rude comments, provided by Tom Tregenza of the University of Exeter in Cornwall (Tregenza 2009), include the following:

- The manuscript reads as though it were written in a couple of days.
- The figures presented are absolutely useless in their current form.
- It must have taken the authors considerable courage to venture into an area which is clearly not their own.

Hence, it is important to be prepared for rejection, hostile comments and/or lists of objections that will be time-consuming and tedious to address. You will then be psychologically ready to do so calmly and effectively.

All that then remains to be done is to wait for the editor's decision.

6.4 After Receiving the Editor's Decision

The most appropriate course of action having received a message from a journal's editor depends on whether he/she has decided to accept the paper without revision, accept it with minor revisions, accept it with major revisions or reject it outright. Therefore, let us consider the best responses to each of these kinds of decision.

6.4.1 Acceptance Without Revision

These decisions are very rare, and they are (of course) by far the easiest to deal with. Firstly, a brief message should be sent to the editor, such as:

Dear Sir/Madam

We thank you and the referees for your careful reading of our manuscript and we are delighted that you have accepted our paper.

Yours Faithfully

The next step is to have a party, take your colleagues out to dinner, go fishing, tackle your ongoing experiments in a happy state, or celebrate in any other manner of your choosing. Having done so, the referees' comments should be read *carefully*. In addition to recommending acceptance, they may have made valuable suggestions that could provide the basis for further interesting experiments, or even a grant proposal.

Furthermore, if you have retained all the relevant materials, samples and reagents, it may be possible to perform suggested experiments, and thus obtain another publication, quickly. In addition, in such cases you could state that you had already planned to perform the suggested experiments, and ask the editor to include a short addition to the *Conclusion(s)* stating, for instance:

Further potentially interesting aspects include xx... (where xx is the referee's suggestion)... We are investigating these aspects in ongoing studies.

This will (a) indicate that you had already thought of the suggested experiments, but merely omitted to mention them, and (b) deter readers from thinking of the experiments and doing them before you.

Finally, if a paper is accepted without revision, it is worth pondering whether it might have been accepted by a higher-impact journal, and if future papers of similar quality should be submitted to such journals.

6.4.2 Acceptance with Minor Revisions

If referees' objections are very minor, for instance, that a few typographical errors need to be corrected, they should be made and the paper should be re-submitted immediately, with a letter thanking the editor and referees for their careful consideration of the paper. The editor will then feel well disposed toward the paper, and having no grounds to reject it or raise any further objections, he or she is likely to accept it rapidly. Other minor objections include various linguistic points. For example, some referees or editors may insist that constructions such as *The results suggest...* should be changed to *We conclude...*, whereas others (who dislike the personal voice) may insist the opposite. In such cases it is rarely worth arguing, since the sense of the sentences concerned will not be affected, and it is better to make the changes and tell the editor that all of the suggested changes have been made.

It should be noted here that many editors and referees have 'pet hates' (minor annoyances that an individual finds particularly irritating, often with little basis) for various terms or grammatical constructions, and it is often impossible to change their opinion about them, even if you present overwhelming evidence that alternatives are equally valid, or better.

The letter should start with some variant of the following:

> Dear Sir/Madam
>
> We thank you and the referees for your careful reading of our manuscript and we are pleased that you have accepted our paper with minor revision.

All of the referees' comments must then be responded to sequentially, and responses should be numbered in a covering letter under headings such as *Reviewer 1: Comment 1*, followed by *Response*, then *Comment 2* followed by *Response* and so on.

This helps the editor and referees to check that you have addressed all of the concerns, and avoid irritating them. Itemizing referees' comments also helps to ensure that you understand their objections, and how many specific points they have made. Sometimes, several points may be raised in a single 'comment'. If so, each of them should be addressed, perhaps under separate sub-headings; *Comments 1.1, 1.2, 1.3* and so on.

It should be noted that some comments may simply be compliments, such as another quoted by Tregenza (2009):

> The reviewers and the Associate Editor have recommended publication and... it seems your paper is close to perfect. Although referee 1 had some hallucinations about missing words (but failed to say where), referee 2 rose to the challenge and has managed to find a few minor flaws.

If so, they should be repeated in the covering letter, with a note thanking the referee for them (this is important because it reminds the journal's editor that the paper has recognized strengths). Briley (2009) also recommends sending a modified manuscript with the changes highlighted, perhaps in yellow, *and* a modified manuscript without highlighting that the editor can (hopefully) send directly to the typesetter.

6.4.3 Acceptance with Major Revisions

If serious objections have been raised, the first response on receiving the message from the editor is likely to be annoyance and anger; a hostile reaction to a study that may have taken months or years to complete is never welcome. Thus, when receiving responses from journals with hostile comments and demands for major revisions, some authors, for example, authors mentioned but not named by Williams (2004) and Martin (2008; not the co-author of this book), recommend waiting for several days before tackling the objections, in order to do so calmly and effectively. As outlined earlier, we recommend being prepared for rejection, and tackling the objections a little earlier than this, because preparing responses may be time consuming and the paper may need to be re-submitted quite quickly. A recommendation by Martin (2008) that we do agree with is *not* reading your initial text first, because that cements the previous approach. Instead, he advocates going through the referees' and editor's comments sequentially, and *then* reading through the paper, and improving the expressions and narrative flow.

Usually, a demand for major revision will be accompanied by a list of objections from two or three referees, some minor and others that are more serious. Regardless of how serious or minor they are, it is essential to read them *thoroughly*, preferably several times, to make sure you understand them fully (if possible, sometimes comments are incomprehensible, see later). On detailed inspection, the required revisions may range from requests to correct typographical errors, through re-writing one or more passages of text and/or changing some figures to demands for complex experiments to be performed. In extreme cases it may even be impossible to meet all of the demands. Thus, the first thing to consider is whether it is worth trying to meet them. There are several factors to consider in this respect, most importantly:

- the tone of the letter
- the length of time addressing the concerns will take
- the time and funds available to address them
- the importance of publishing the paper (e.g., to support a grant application or to meet requirements for a PhD)
- your experience; if you have not written a paper before, or only a few, you may be surprised (and dismayed) about the numbers of objections that are often raised, and think that it will be impossible to address them all, but it may be possible to deal with them reasonably quickly if the task is approached calmly and systematically.

The tone of the letter is an important factor since it may be quite encouraging, saying, for instance, that if certain sections of the paper are re-written and/or some controls that will be easy to analyze are included, the journal will be pleased to publish the paper. Alternatively, it may be quite harsh, stating that *if* extensive additional experiments are performed that corroborate the results, *and* large sections of the paper are re-written, *then* it will be re-considered, but worded in such way that the chances of acceptance seem remote.

Re-writing passages may take several days, but if that is all that is required, it should at least be feasible, even if the editor demands a complete re-write and your linguistic skills are not major strengths. In such cases, there are strong incentives for starting the revisions quickly, and re-submitting the paper as soon as possible, after sending it to a good professional editor for checking following the re-write, if necessary.

At the other extreme, if substantial further work is demanded, and the tone of the letter is very discouraging, strenuous efforts may be needed to meet the requirements, and even then re-submission may not be successful. Hence, in some cases there may be little point in striving to fulfil referees' demands. Persistence is sometimes rewarded, and papers that were initially severely criticized have often been published (Campanario 1996), but when chances of publication by the target journal seem remote, submission to another journal, or simply accepting that the study has such serious flaws that it is unlikely to be accepted by any journal, can be considered (see later).

Usually, when editors ask for major revisions, the requirements fall between these extremes, and the objections *can* be addressed with varying degrees of time, patience, further experiments and/or re-writing, if you decide the additional effort is warranted. If you do decide to re-submit the paper, the next step is to group the referees' comments concerning specific aspects of the paper.

In some cases each of three referees may have raised (say) 20 objections, giving a total apparent list of 60 objections to tackle. However, on closer inspection, 12 of the objections that they all raise may be similar quibbles about linguistic points (e.g., typographical errors) that can be easily addressed, thereby reducing the list of objections that remain to be tackled to 24. There may also be very strong overlaps between the other objections they have raised, some of which may be very straightforward to deal with, for example, two or three of them may say that several references you have not included should be cited, and/or that the same paragraphs should be re-written. If you have no objections to citing the mentioned references and re-writing the paragraphs, this should be done. At this point, or any point hereafter, you can start to draft a letter, beginning (as mentioned earlier) with something like:

Dear Sir/Madam

We thank you and the referees for your careful reading of our manuscript and we are pleased that you have accepted our paper provided revisions are made, which we have done, as detailed below.

Again, it should continue by sequentially responding to referees' comments under headings such as *Reviewer 1: Comment 1*, followed by *Response*, *Comment 2* followed by *Response* and so on, mentioning the changes you have made, and where they have been made. However, having addressed a point (say) in your *Response* to *Reviewer 1: Comment 5,* you can then simply write *See our response to Reviewer 1's Comment 5*, in your responses to corresponding comments by the other referee(s).

Having identified, grouped and dealt with simple objections, there may only be a very few serious objections to address. The way in which they should be tackled depends on the following:

- the referees are right or wrong
- the referees are unanimous or disagree with each other
- the referees' comments are clear, open to interpretation or incomprehensible
- the paper would be excessively long (or short) if objections were addressed
- additional work that the referees suggest is feasible.

Let us consider each of these scenarios.

6.4.3.1 The Referee Is Right

Referees are usually experts in the field covered by papers they review. Thus, they often make very good suggestions that will substantially improve the paper. For instance, they may draw your attention to an important reference that you should have included, points you should have raised, statistical tests you should have applied, or intriguing alternative interpretations of your data. If so, there are two possible kinds of responses:

- Simply thank the referee for drawing your attention to the omission or other interpretation, incorporate the relevant information in the paper, and tell the editor you have done so.
- If you are embarrassed about omitting something that clearly should have been included, again thank the referee for drawing your attention to the omission or other interpretation, incorporate the relevant information in the paper, and tell the editor you have done so, adding that you meant to include it in the original paper, but forgot.

As mentioned before, referees may also correctly suggest controls that should be included, or other experimental work that needs to be done, for instance, to validate your results. If the suggested experiments can be feasibly done, and you decide that publishing the paper warrants the additional work, they should be done as soon as possible (hopefully yielding desirable results), and the paper should be amended accordingly, provided this does not make the paper excessively long (see later). At this point, having relevant materials, samples and reagents to hand is extremely valuable.

6.4.3.2 The Referee Is Wrong

Objections that are wrong should be *courteously* disputed or (often better) acknowledged. Statements like *We totally disagree* or *The referee clearly misunderstands...* should be avoided, since they will alienate the referee. There are two main strategies that can be adopted:

- State, in your covering letter, why you disagree with the referee, coherently and preferably with supporting references (mentioning any supporting data that you did not previously present, data presented in another section of the paper that the referee has overlooked and/or any further reasons for focusing on the selected phenomena or using the selected methodology that you may have thought of since submitting the paper).
- Use phrases in the covering letter such as *The referee raises an interesting point, and highlights a passage in the text that required clarification.* The reasons why you think your interpretation is correct can then be outlined in the letter before saying, for instance, *To clarify this point, we have amended the text* (followed by the page and lines where the changes have been made). Then, if you think the referee is wrong, but his/her interpretation is at least theoretically plausible, you can briefly mention the point he/she raised and the reasons you believe your interpretation is more valid, at an appropriate place in the paper.

The first approach should be avoided, if possible, since it clearly implies that the referee was wrong, forcing the editor of the journal to decide which of the two interpretations is most likely. This is highly undesirable because he/she may side with the referee, and even if he/she agrees with you, the referee will be annoyed if there is a further round of review. The latter approach is preferable, in these respects, since it acknowledges that the referee *may* be right (even if you are sure he/she is wrong). Furthermore, other readers may also come to the same conclusion as the referee, providing a further reason for clarifying why you think your interpretation is valid in the text.

The second strategy cannot always be applied, since (for instance) a lengthy paragraph may be needed to describe and refute the referee's view. This may cause word limits to be exceeded, or it may cause a section to be unbalanced, that is, if it is included, half the *Introduction* may be focused on a relatively minor part of the *rationale*. However, points raised can often be incorporated in a single sentence. For instance, a referee may argue that cited references do not demonstrate that cruciferoid levels are modulated by temperature and light because observed variations may have been due to variations in unmeasured variables, such as nutrient levels. If so, you can add *Although the reported results strongly suggest that temperature and light modulate cruciferoid levels, it is possible that other factors that were not examined in the cited studies, for example, differences in nutrient availability, were responsible for the observed variations*. In other cases, it is not even essential to change the meaning of the amended passages, sometimes merely re-writing them slightly to show that you have not ignored the referees' remarks is sufficient.

In further cases, a referee may simply prefer a technique other than one used in the study. For instance, a paper we know of was initially rejected because plant hormones were measured in the described study by gas chromatography-mass spectrometry (GC-MS), but the referee thought that enzyme-linked immunoassays (ELISAs) provided more reliable data. However, the authors were able to counter this successfully by pointing out that although ELISAs could have been used, GC-MS had often been used for the purpose, and there were published indications that it provided more robust measurements. They also incorporated a further sentence justifying their use of GC-MS in the text.

It should always be remembered that most referees spend considerable time, without pay, refereeing papers, which is one reason that their comments are very rude sometimes. Hence, even when you receive long lists of comments, some of which may be quite hostile, it is usually counter-productive to alienate the referees and editor by being rude, arrogant or pompous, however irritating or wrong you may think some objections are. Williams (2004) also recommends asking someone else to read your responses before sending them to the editor, which is very sensible, and Martin (2008) recommends paying most attention to the *editor's* suggestions. If your revisions are thorough, the editor may decide to accept the paper without sending it out to reviewers again, or he/she may regard your work more favorably when he/she receives their comments.

6.4.3.3 The Referees Are Unanimous

When all the referees raise a similar serious objection, the situation is at least clear cut: It *must* be addressed if the paper is to be accepted, so you simply need to decide if the suggested experiments can be feasibly done and if publishing the paper warrants the additional work.

6.4.3.4 The Referees Disagree with Each Other

Dealing with cases where two referees strongly disagree with one another may at first seem difficult. However, this is really very helpful. If one supports your methodology or interpretations, while the other is highly critical, you can use the arguments presented by the former to support your case. For example, you can add in the text something like *An alternative, possible interpretation of our results is that* (followed by the dissenting referee's interpretation). *However, it seems more likely that* (followed by your interpretation). Similarly, if the referees disagree with both you and each other, you could add something like *Alternative possible interpretations of our results are that* (followed by both dissenting referees' views). *However, it seems more likely that* (followed by your interpretation). The editor should then be told that you have incorporated the alternative interpretations in the text, mentioning where they have been inserted. Or, if on reflection they all seem to be valid possible interpretations, you could amend the paper by saying *There are several possible explanations for these findings, including* (followed by all three interpretations) and again tell the editor that you have made this change.

6.4.3.5 The Referees' Comments Are Clear

When referees' objections are clear, they may be difficult to address, but at least there is no doubt about the issues that must be tackled, in which case they must be addressed in one of the ways outlined earlier.

6.4.3.6 The Referees' Comments Are Open to Interpretation or Incomprehensible

Sometimes referees' objections and/or editor's remarks can be difficult to interpret. As an example, Williams (2004) cites phrases such as *We cannot accept your paper in its current form, but if you do decide to resubmit, then we would only consider a substantial revision*. This may seem close to an outright rejection, but in reality it may simply be an invitation to resubmit (e.g., the editor may be merely trying to ensure that you address the referees' comments seriously). If you are unsure about how to interpret such responses, Williams recommends either asking an experienced colleague for their advice (or, better, someone who works as a referee for the journal) or writing back to the editor to ask for clarification.

Addressing a referee's comments that are open to interpretation or incomprehensible may also seem difficult at first, but such comments may be very helpful. First, you should check that colleagues also cannot understand them (i.e., that the failure to see what the referee is trying to say is his/her fault rather than yours). If they agree with you, as Briley (2009) suggests, you can then use a politician's response, that is, interpret them in a way that is relatively easy to address, prefacing your response with a phrase such as *If I understand the referee correctly....* He notes that the editor will probably accept your interpretation of the question (and probably will also have little idea of what the referee was trying to say). Similarly, Martin (2008) notes that if you can draft changes that seem plausible, they are likely to be sufficient for the editor. However, you should always think carefully about what the referee is trying to say first, since with sufficient concentration it may become apparent.

6.4.3.7 The Paper Would Be Too Long (or Short) If Objections Were Addressed

In some cases, referees and the journal's editors may insist that lengthy passages should be added to a paper, but other passages should be cut, so overall there will be little change to the paper's overall length. If so, you may not want to cut the latter passages, but at least this may lead to the paper being accepted (usually the main objective). Furthermore, it may be possible to submit the deleted information as a short communication to another (probably lower-impact) journal, thus yielding two publications instead of one. In other cases, referees may request the addition of more text, or data from further experiments, which would cause word limits to be greatly exceeded. This can also be very helpful, since it is possible to point this out to the editor and suggest that the information be presented in two papers, again yielding two publications (possibly two full papers) rather than just one.

In other cases, editors may demand the deletion of large sections, causing the paper to be so short that it should be published as a short communication. If so, you may decide to accept the offer and at least obtain a publication. Furthermore, again it may be possible to submit the information in the deleted sections to another (probably lower-impact) journal as another short communication. Alternatively, if you strongly want to retain all of the information, it may be necessary to submit the paper to another journal, following considerations discussed later.

6.4.3.8 Suggested Work Is Not Feasible

In some cases, severe criticisms by a referee may be correct, but requested changes are: *not feasible* or *idealistic* (e.g., running suggested simulations may take years of computer time, or no mutants may be available for suggested analyses) or *beyond the scope of your study*. For instance, a referee may say that attempts should have been made to elucidate the physiological roles of compounds you investigated, which would certainly be interesting, but you focused on the first steps toward such elucidation (identifying and measuring them), and even this was a major undertaking. Briley (2009) notes that responses to such comments can be prefaced with phrases such as *The referee is theoretically correct, but...* followed by explanations (with references, if appropriate) why it would not be possible or feasible to follow the referee's suggestions. He also notes that you should remember that you are writing your replies to the *editor*, not the referees, that he or she will be very busy, and he/she has probably, at best, scanned your manuscript and is probably not an expert in the area, in contrast to you and (at least theoretically) the referees. Hence, most editors he knows will usually accept a modified manuscript from an author who has made an effort to answer at least half of the comments well, and has given reasonable arguments for the rest, concluding that authors should not despair if they have trouble addressing one or two points.

6.4.4 Outright Rejection

Usually letters telling authors that a paper has been rejected are quite short and offer very little opportunity to resubmit. The main reasons for rejections are the same as those that prompt referees' objections: unsuitability for the journal; lack of sufficient novelty, interest or importance; or 'lethal' methodological concerns. As the fate of the 'citation classic' mentioned earlier shows, it is occasionally possible to show that the subject of the paper *is* relevant, despite a referee's objections. Similarly, if the 'lethal' objection is that the true causes of variations in phenomena you have investigated may have been factors that you failed to consider, it *may* be possible to acquire data that allow the effects of these factors to be assessed. If so, you could incorporate further analysis in your paper and re-submit, with explanation that the 'lethal' concern has been addressed. In addition, if you believe that the referees' comments are grossly unfair or simply wrong, you *can* appeal to the editor and ask for new referees. However, such appeals are rarely successful, and the chances of successful re-submission following outright rejections are generally very slim.

A better option, usually, is to submit elsewhere, after reading and assessing the referees' comments as thoroughly and objectively as possible (regarding referees' objections as free expert advice), since as mentioned before addressing the comments may substantially improve the paper (especially if the referees' opinions are unanimous) and the revised paper may be sent to the same referees even if you send it to another journal. In addition, if the main objections focus on a particular part of the study, and the rest of the investigation could be presented in a shorter paper in another journal, the chances of success can be boosted by deleting all sections of the paper related to that part.

If you decide to submit the paper elsewhere, with or without amendments, the next step is to choose another journal. Normally, in such cases a journal with a lower impact rating is chosen, or one with a higher acceptance rate, on the grounds that its selection criteria will be less rigorous. However, Bornmann et al. (2009) found that papers rejected by *Angewandte Chemie International Edition* that prompted large numbers of referees' objections in categories designated *Relevance of contribution* and *Design/Conception* were unlikely to be published later in a similarly ranked journal. In contrast, numbers of objections in categories *Writing/Presentation*, *Discussion of results*, *Method/Statistics* and *Reference to the literature and documentation* were not significantly correlated to the probability that a rejected manuscript would be subsequently published in *either* a low- or high-impact journal.

Therefore, the nature of the referees' objections also needs to be considered. These findings indicate that if objections centre on the writing, discussion, statistical analysis or references, fairly minor amendments to a paper may be sufficient to secure acceptance by another journal, even one of similar ranking to the first journal. However, if they concern the relevance or design of the study, you may have to repeat some of the investigation (with a different or extended design) and/or think of stronger justifications for your work and the design of the investigation and incorporate them in the revised paper. Furthermore, there are many anomalous responses. For example, a paper by an author we know of was rejected by a moderately ranked journal, and then by a less highly ranked journal. In a fit of pique he sent it to *Nature*, which promptly accepted it.

References

Bishop CT (1984) How to edit a scientific journal. ISI Pr, Philadelphia, PA

Bornmann L, Weymuth C, Daniel H-D (2009) A content analysis of referees' comments: how do comments on manuscripts rejected by a high-impact journal and later published in either a low- or high-impact journal differ? Scientometrics 83:493–506. doi:10.1007/s11192-009-0011-4

Briley M (2009) Replying to referees. http://www.psy-world.com/replying_to_referees.htm. Accessed 11 Aug 2010

Campanario JM (1996) Have referees rejected some of the most-cited articles of all times? J Am Soc Inf Sci 47(4):302–310

Gurr MI (1990) Editorial, the journal's peer reviewing system. Br J Nutr 63:410–407

Martin B (2008) Surviving referees' reports. J Sch Publishing 39(3):307–311

Relman AS, Angell, M (1989) How good is peer review? Letter 21 Sept. N Engl J Med 321: 827–829

Tregenza T (2009) Collaborators. http://www.selfishgene.org/Tom/Collab.htm. Accessed 3 Dec 2010

Williams HC (2004) How to reply to referees' comments when submitting manuscripts for publication. J Am Acad Dermatologists 51(1):79–83. doi:10.1016/j.jaad.2004.01.049

Chapter 7
Other Kinds of Written Scientific Communication

This book has focused on writing peer-reviewed papers, since they are key elements of the primary literature, and every scientist who wants a successful academic career needs to write such papers. However, scientists communicate with each other in diverse ways nowadays, ranging from brief emails, letters, text messages, blogs, Facebook profiles and Twitter 'tweets', through press releases, newsletters, oral communications and reports for funding bodies or investors, to grant proposals, short communications, full papers, theses, posters and reviews. Sometimes, too, a scientist's sub-conscious mind may send important messages to his or her conscious mind (notably, e.g., the messages Kekulé received, see later). These communications (at least the conscious ones) should be composed carefully since they can have profound effects on scientists' careers, some helpful and some highly disadvantageous. Therefore, these kinds of communications are considered in this chapter.

Famously, in a speech to the German Chemical Society (1890) Kekulé claimed to have had two 'daydreams' (Kekulé 1958). In one, atoms gamboled before his eyes then, frequently, two smaller atoms united to form a pair, larger ones embraced the two smaller ones, still larger ones kept hold of three or four of the smaller ones, and formed a chain, dragging the smaller ones after them. This was the origin of his Structural Theory (see Rocke 1981). In the other, he saw a serpent seize its own tail, and realized that this explained the structure of benzene, the ring structure accounting for its apparent violation of chemical valency rules. Kekulé's true mental states during these 'daydreams' have been disputed for various reasons (which do not concern us here). Unfortunately, whatever they were we are unable to give you any advice about entering such states in order to obtain brilliant insights, therefore, we will turn our attention to other forms of communication that can be more readily controlled.

7.1 Electronic Communications

Email and messages posted to social networking websites are critical media nowadays for keeping in touch socially, maintaining relationships (good and bad) with other members of scientific communities (e.g., actual and potential collaborators, referees, members of funding council's boards and editors of journals),

J. Blackwell, J. Martin, *A Scientific Approach to Scientific Writing*,
DOI 10.1007/978-1-4419-9788-3_7,

and conducting diverse kinds of transactions. They are highly valuable because they allow virtually instantaneous communication with large numbers of people. However, this is also a major weakness, since it means messages can be written and transmitted in haste, unlike letters (which can be torn up before reaching a post box, if the writer realizes that the contents are insulting, libelous or foolish). Furthermore, emails cannot be recalled, and although postings to networking sites can be subsequently amended, many people may already have seen their contents and distributed them widely via the internet.

Hence, emails and electronic postings should be written *more* carefully than letters, rather than less carefully, and in a calm rather than an angry or agitated state. It is important to punctuate them properly, and to refrain from being rude, partly because they may well reach a far wider audience than you originally intended. For instance, it may be very satisfying after a grant proposal has been rejected to write an offensive diatribe against the head of the funding council that rejected it. However, in a few days' time you may well be dismayed to see messages on networking sites such as:

> Have a look at the great description of Xxxxx written by Jim... A pompous, arrogant idiot. I knew him when he was a post-grad. He didn't understand basic science then, his understanding hasn't improved with age, and he has less idea about a good project than my half-mad cat.

Even if a message is unlikely to reach a wider audience, it is generally essential to avoid being rude in such communications, even unintentionally, and to write the kind of message that you would like to receive if you were in the recipient's position. For example, if you have submitted a paper to a journal and you have not had an acknowledgement for a week, and hence you are worried that the editor is unnecessarily delaying processing it, it is unwise to send a message saying

> Dear Sir/Madam
>
> We sent a paper to the *Journal of Bananas* entitled *Mature bananas are yellow and bent* a week ago, but have not had an acknowledgement. This is poor service. Have you received it?

Such a message is unlikely to make the editor feel well disposed to either you or your paper. Thus, a better option is something like

> Dear Sir/Madam
>
> We are concerned because we sent a paper to the *Journal of Bananas* entitled *Mature bananas are yellow and bent* a week ago, but have not had an acknowledgement. We are worried that it may not have reached you, therefore please let us know if you have received it so we can send another copy if necessary.

Similarly, if you want a potential collaborator to send samples for analysis as quickly as possible, we would not send a message stating

Dear Bob

Please send the samples AS SOON AS POSSIBLE!!! We want to analyze them very soon and we are just WAITING ON YOU!!!

Jim

Such a message may seem very rude and impatient to the recipient, even if you had no intention to be rude and merely wanted to let him know that you were eager to start work. Indeed, Bob may decide to send the samples to another, less impertinent, potential collaborator instead. Therefore, a much better alternative is something like:

Dear Bob

We are very eager to start analyzing the samples we discussed, because we think the acquired data could fill major gaps in our understanding and lead to a very high profile publication. Therefore, we would be very pleased if you could send them as soon as you have time so we can start work on them.

Hoping you and your family are all well

Jim

Adding some kind of short personal comment (e.g., *Hoping you and your family are all well* or *I hope you are finding time to get out of the lab now and then* or *Have you caught any good fish lately?*) to messages to collaborators, colleagues, and anyone with whom you have anything other than a purely business or professional relationship, is often good practice, because it helps create a friendly attitude and prevents messages from being cold, impersonal and terse. In addition, it helps prevent rudeness creeping into other parts of the message. For instance, it is difficult to write a message such as:

Dear Bob

Please send the samples straightaway, we need them now and you are holding things up.

Hoping you are fit and well

Jim

Of course, there are occasions when rudeness *is* appropriate, for example, if Bob has been promising to send samples for months, especially if you are his supervisor. In such cases, more direct and even threatening messages may be required, such as

Bob

Send the samples NOW, or look for another job.

Jim

However, generally speaking, it is best to adopt a polite, professional, friendly (but not over-friendly) tone in electronic communications. Numerous authors offer detailed advice on email etiquette (e.g., the Institute of Education undated), and readers who are not familiar with such advice are strongly recommended to read (and *apply*) the guidelines from a reputable source, in order to foster good relationships with potentially critical people.

7.2 Other Communications

In a short book, it is impossible to describe, in detail, approaches for preparing and presenting all of the diverse kinds of scientific communications. However, the principles and strategies outlined earlier for drafting and writing peer-reviewed papers can be applied to all of them. Notably, they should be:

- Clear, concise and cover all of the main points.
- Coherent, using appropriate sub-headings and/or flow charts if need be to structure your thoughts, information and arguments (even if you do not include the sub-headings in the final text).
- Jargon free, or more strictly, the scientific terms used should be appropriate for the audience, for example, a press release should contain as few technical terms and as many standard words, as possible, while this is much less of a concern for an address to eminent scientists in your field.

We will conclude this guide with a brief section on reviews, focusing on a series of reviews written by one of the authors, during the course of which the reference mapping and some of the argument-structuring strategies originated.

7.3 Reviews

As nearly all readers will be aware, reviews differ from standard papers in that (as the term suggests) they review, or summarize, knowledge of the subject addressed. Hence, they consider, and critically appraise, relevant parts of many previous studies regarding the subject of interest rather than describing a single study. However, like ordinary papers, they have a standard format that provides a convenient overall template, consisting of an *Introduction*, *Sections describing each aspect of interest*, usually (but not always) a *General Discussion*, *Conclusions* and *References* (both essential and illustrative).

Occasionally, the authors of this book have been commissioned to write reviews about subjects that we know little about. At first, this may seem very challenging. However, it is not impossible if the task is approached systematically. The first key issue to consider is *Have there been any previous reviews?*

If so, they should provide at least most of the information required regarding knowledge up to their publication dates. Thus, the information they contain can be summarized, and they will probably provide suitable sub-headings for each section of the planned review (with a little modification to avoid accusations of plagiarism). In addition, they will provide key words that can be used to search for literature on each aspect to cover that has been published since the previous reviews. If there have been no previous reviews, then a few recent papers on the subject must first be scanned, and they should yield all of the key words needed to identify other relevant papers that need to be cited.

r, a number of papers on measurements of process parameters will be discussed in
nore detail, beginning with a few on sugar polarization (pol) measurements. The
Snoad (2009a), describes use of an electrical conductivity (EC) method (applied
ix solutions) that has been successfully used at Mulgrave Central Mill in Australia
seasons. During this time the standard deviation of the difference between results
with the EC and conventional methods has been 0.093 Z. The method is also
it requires little analytical skill and its use (in conjunction with an EC method
ssing pan product true purity) has allowed significant reductions in laboratory
er, without reductions in analytical output. On the other hand, according to the
turers at least, those in 'pursuit of perfect polarimetry' need to look no further than
n Paar Optotec Sucroloyser System, consisting of a Sucromat VIS/NIR saccha-
Abbemat HP refractometer, EasyFilt and Sugar Lab Software (Groves and Orlando
his system measures pol by NIRS (Near Infra-Red Spectroscopy), following filtra-
ch as the cited authors note is more environmentally friendly and less hazardous
tives than conventional pol measurements following lead acetate-driven precipita-
codel and Hoareau (2009) also prefer filtration followed by NIR determination of
e same reasons. However, they also note that the measurement of pol per se has
s since various compounds in cane, in addition to sucrose, affect optical rotation
nents (including glucose, fructose and other saccharides). Further, they report, pol
tained using both the lead acetate and NIR-based methods yield results that deviate
h-Performance Liquid Chromatography results, and the deviations vary depending
mple (extract/juice/bagasse/molasses) assayed.

e, NIRS has become increasingly popular for measurements of huge ranges of
rs, and in previous years quite large numbers of papers on applications of NIRS
d. This year there have been slightly fewer, but another one will be mentioned. This
Simpson and Oxley (2009), concerns the use of NIRS for the routine (weekly)
f molasses (pol, Brix, dry solids, conductivity ash, fructose, glucose, sucrose and
rity difference) and mixed juice (pol, Brix, conductivity ash, fructose, glucose and
y the South African Sugar Milling Research Institute. The system was set up for
ghput analysis, with minimal time requirements, and the results were compared
obtained by traditional laboratory methods, for week-by-week analysis of the
capabilities of the NIRS. For molasses samples, the NIRS-predictions showed
ents over previously published data.

t to his earlier-mentioned endorsement of EC for estimating pol, Snoad (2009b) is
nced that it is so suitable (despite being simple, quick, relatively cheap and widely
monitoring contamination of boiler feed water by process materials (which can
blems related to boiler priming, corrosion, scaling and water treatment costs). He
most material in mill products causes little or no change in EC, thus relatively
entrations of contaminants may accumulate before they are detected by EC probes.
most alternative options require complex equipment, reagents, calibration and
lysis times. A possible solution he discusses may be to use UV photometry.

written the sugar technology review, each year it was sent to several
eading to ensure that there were no obvious errors and to acquire feed-
sections that should be amended. The suggested corrections were usually
Thus, these reports show that reviews, and other complex texts, can be
with coherent, fully referenced arguments, *even when one has little
of the subject*, provided that the exercise is approached systematically.

The next essential issue is to formulate sub-headings for each section of the planned review. If the subject can be broken down into chronological steps, this is straightforward. For example, if the life cycle of a plant is being reviewed, sub-headings could include *Embryogenesis*, *Seed germination*, *Vegetative development* and *Flowering*. If the subject cannot be readily divided into chronological steps, it should be divided into convenient topics. For example, if the roles of hormone receptors in Anon's Disease are being reviewed, suitable sub-headings could include *Symptoms and pathogenesis of Anon's Disease*, *Hypotheses postulated to explain its pathogenesis* and *Interactions between HGFF, HGFF receptors and HGFFs kinase*.

Often, even when the reviewed phenomena have a strong chronological dimension, there will be some aspects that cannot be readily included in chronological sub-sections, or only included with tedious repetition that disrupts the narrative flow. In such cases, these aspects can be addressed in separate sub-sections after describing the chronological sequence. For example, a review of plant development may include passages on *Roles of plant hormones* and *Effects of environmental variables*, which *could* be incorporated in each of the sections, since these factors affect every stage of plant development. However, this may make the review too long, if word limits are tight, and in such cases, it may be better to include separate sections in which a few of their effects are selectively considered to illustrate their importance, after the chronologically arranged sections.

Having constructed sub-headings for each section, the papers that need to be cited (identified as outlined in previous chapters) can be assigned to relevant sections, and grouped under further sub-headings according to the topics covered within the sections. For example, for an *Effects of plant hormones* section, papers that illustrate effects of each of the main classes of plant hormones could be grouped under sub-sub-headings (which may be subsequently deleted), such as *Effects of auxins* and *Effects of cytokinins*.

The papers should then be scanned, and the information from each paper that the author wishes to mention should be very briefly summarized, under the previously constructed sub-sub-headings. The review will then already be structured at the overall, section and sub-section levels. In addition, the papers that need to be cited, and the precise points that they need to be cited, will have been identified. Hence, all that then remains to be done is to link the information using appropriate conjunctions to provide narrative flow, present hypotheses (with supporting and conflicting information), and structure the paragraphs, then insert the references at the appropriate points.

It should be noted that if a review covers a very broad subject, it is impossible to be expert in every aspect. Thus, the most challenging stage of writing a broad review of a subject is addressing, and critically appraising, conflicting views regarding aspects that one knows relatively little about. Fortunately, however, authors with opposing views nearly always provide critiques of each other's work, so it is quite straightforward to access conflicting hypotheses and present alternative views regarding any issue. Indeed, if a hypothesis is particularly controversial, there will be 'chatter' about it very quickly on the internet, which can be mined for conflicting interpretations, even if there are no published critiques.

To illustrate this process, one of us (JB) used to write ca. 18,000-word annual reviews of developments in sugar-processing technology. This was quite easy even though it was far from JB's fields of core competence, despite acquiring a smattering of knowledge since two of several titles he edited for a few years while working for International Media Ltd. (subsequently part of Informa) dealt with sugarcane and beet agronomy, processing and economics. Fortunately, sugar processing can be easily broken down to a series of chronological steps, as follows:

- harvesting and transport of the sugarcane or beet
- extraction (milling or diffusion for sugar cane; slicing, cleaning, dewatering and diffusion for beet)
- purification (clarification, decolorization and demineralization)
- heating, evaporation and crystallization
- centrifugation, drying, cooling and storage
- waste disposal and emission control.

In addition, there are several aspects that do not fit into this scheme, but can be written about after the chronological steps:

- mill sanitation
- energy and efficiency
- measurement of process parameters
- process and quality control
- miscellaneous, that is, there were usually a few papers that did not readily fit into any of the categories mentioned earlier.

These categories provided very convenient sub-headings. The next step was to scan the literature to identify papers that had been published on sugar processing since the previous annual review, and assign ('map') them to the relevant sections and sub-sections, for example, perhaps first to evaporation, and then to specific kinds of evaporators, which are used to concentrate purified sugar-containing solutions to concentrations at which the sugar can be readily crystallized.

Having done this, there were usually just a few papers per group pertaining to each sub-section of the review. For the 2010 review, the papers identified for part of the *Measurement of process parameters* section were:

Corcodel L, Hoareau W (2009) Influence of polarisation measurement on the establishing of material balance in sugar manufacture. Paper presented at the 26th Association Andrew van Hook Symposium, Reims, France

Groves D, Orlando L (2009) Pursuit of perfect polarimetry. Paper presented at the South African Sugar Technologists' Association Congress, Mount Edgecombe, RSA

Simpson R, Oxley J (2009) Routine analysis of molasses and mixed juice by NIR spectroscopy. ISJ 111:387–402

Snoad JD (2009a) Assessment of sugar polariza
ity methods. ISJ 111:306–312

Snoad JD (2009b) Preliminary investigation i
method as an option for detecting contamina
Technol 31:451–459

The next step was to read the *Abstracts* of the pa
to obtain most of the information required), and dec
novel information to cite them (except in irritating c
uninformative so the whole paper had to be read). T
each paper was jotted down in a sentence, if possibl
a very short paragraph. Finally, the sub-sections v
were inserted in appropriate places, as described ea

As also mentioned before, the most challenging
subject is critically appraising conflicting views re
the author's fields of competence. This essentially
the sugar technology reviews. Indeed, it would ha
in all of the aspects covered since sugar processin
alia) harvesting equipment; transport logistics; mi
tion; various approaches for 'clarification' (the c
contaminants); diverse strategies to remove colore
ing, concentrating and crystallizing sugar solutio
from residual ('molasses') streams; methods for e
desired sizes; several types of systems for drying
ing sugar; use of antibiotics to control microbial
break down microbial products (notably dextran);
removing contaminants from boiler emissions, a
for removing pollutants from liquid waste strea
by-product gases); optimization of energy and ste
energy cogeneration or other uses of solid wastes
process parameters and equally numerous syste
processes.

Clearly, it is not feasible to be an expert in al
the lack of expert knowledge *did not matter*, si
sources, provided all of the required informatio
was relatively straightforward to paraphrase rele
be cited, after grouping them in appropriate se
them to present the information coherently and
the author's knowledge of the subject. Using thi
maintained, and coherent arguments can be con
knowledge of complex subjects. For instance, p
lier were used, with paraphrasing, in the 2010
passage (very slightly modified for consistency
book):

Howeve
a little
first, by
to 27°B
for thre
obtaine
simple,
for asse
manpow
manufa
the Ant
rimeter,
2009). T
tion, wh
for oper
tion. Co
pol for t
limitatio
measure
values o
from Hi
on the sa

Of cours
paramete
were cite
paper, by
analysis
a target p
sucrose)
high-thro
to results
predictive
improven

In contras
less convi
used) for
cause pro
notes tha
high conc
However,
longer an

Having
experts for
back about
very minor
constructed
knowledge

References

Institute of Education (undated) Email etiquette. University of London, London. http://www.ioe.ac.uk/about/documents/About_Policies/Email_etiquette.pdf. Accessed 22 Aug 2010

Kekulé A (1958) 'August Kekulé and the birth of the structural theory of organic chemistry in 1858' speech to the German chemical society in 1890, translated by Benfey OT. J Chem Educ 35:21–23

Rocke AJ (1981) Kekulé, Butlerov, and the historiography of the theory of chemical structure. Br J Hist Sci 14:27–57

Chapter 8
Summary

Writing a paper is not easy, especially if you are not writing in your first language, but approaching the task methodically can ease the process, by:

- Delimiting the study.
- Writing brief (subsequently expanded) statements describing the rationale and specific objective(s) of the study, what was done, the main findings, novel aspects and limitations (focus) of the analyses, and finally the implications of the findings.
- Using these statements to compose each section of the paper, which should be written clearly and simply, following the target journal's *Instructions for authors*, providing clear figures and tables, where appropriate.

This procedure will help to maximize the chances of your paper being accepted.

However, do not be too dismayed if it is rejected – this does *not* necessarily mean that your work has no merit.

J. Blackwell, J. Martin, *A Scientific Approach to Scientific Writing*,
DOI 10.1007/978-1-4419-9788-3_8, © Springer Science+Business Media, LLC 2011

Subject Index

J. Blackwell, J. Martin, *A Scientific Approach to Scientific Writing*,
DOI 10.1007/978-1-4419-9788-3, © Springer Science+Business Media, LLC 2011